Praise for
The Geography of Thought

"Nisbett boldly . . . challenges the assumption that all people everywhere think the same way."

—*Booklist*

"This outstanding book makes key contributions to education, science, health, business, politics, language, and religion."

—*Library Journal*

"I have long been following Richard Nisbett's groundbreaking work on culture and cognition. After so many fascinating experiments, challenging hypotheses, and passionate debates, it was time for Nisbett to share his ideas and findings with a wider public. *The Geography of Thought* does so superbly!"

—Dan Sperber, author of *Explaining Culture: A Naturalistic Approach*

"An important, research-based challenge to the assumption, widespread among cognitive scientists, that thinking the world over is fundamentally the same."

—Howard Gardner, Harvard University, author of *Frames of Mind: Theories of Multiple Intelligences*

"This is another landmark book by Richard E. Nisbett. Nisbett shows conclusively that laboratory experiments limited to American college students or even individuals from the Western Hemisphere simply cannot provide an adequate understanding of how people, in general, think. The book shows that understanding of how individuals in Eastern cultures think is not just nice, but necessary, if we wish to solve the problems we confront in the world today. We ignore the lessons of this book at our peril."

—Robert J. Sternberg, IBM Professor of Psychology and Education; Director, Center for the Psychology of Abilities, Competencies, and Expertise (PACE Center), Yale University; President, American Psychological Association

"Cultural psychology has come of age and Richard Nisbett's book will surely become one of the canonical texts of this provocative discipline. *The Geography of Thought* challenges a fundamental premise of the Western Enlightenment—the idea that modes of thought are, ought to be, or will become the same wherever you go—east or west, north or south—in the world."

—Richard A. Shweder, anthropologist and William Claude Reavis Professor of Human Development at the University of Chicago

"The cultural differences in cognition, demonstrated in this groundbreaking work, are far more profound and wide-ranging than anybody in the field could have possibly imagined just a decade ago. The findings are surprising for universalists, remarkable for culturalists, and, regardless, they are most thought-provoking for all students of human cognition."

—Shinobu Kitayama, Faculty of Integrated Human Studies, Kyoto University

THE GEOGRAPHY OF THOUGHT

How Asians and Westerners

Think Differently . . . and Why

৵৽৽

RICHARD E. NISBETT

FREE PRESS

NEW YORK LONDON TORONTO SYDNEY

_ƒ_P

FREE PRESS
A Division of Simon & Schuster Inc.
1230 Avenue of the Americas
New York, NY 10020

First Free Press trade paperback edition 2004

FREE PRESS and colophon are trademarks of Simon & Schuster, Inc.

For information regarding special discounts for bulk purchases,
please contact Simon & Schuster Special Sales:
1-800-456-6798 or business@simonandschuster.com

Manufactured in the United States of America

10   9   8   7   6   5   4   3   2   1

The Library of Congress has cataloged the hardcover edition as follows:
Nisbett, Richard E.
The geography of thought : how asians and westerners think differently
. . . and why / Richard E. Nisbett.
p.   cm.
Includes bibliographical references and index.
1. Cognition and culture. 2. East and West. I. Title.
BF311 .N565 2003
153.4—dc21
2002032178
ISBN 0-7432-1646-6
ISBN 0-7432-5535-6 (Pbk)

Illustration credits
Page 23: Matthew Nisbett
Page 216: Jimmy Chan, Oxford University Press

For Matthew
Young man going east

CONTENTS

Acknowledgments / ix
Introduction / xiii

1   The Syllogism and the Tao
    *Philosophy, Science, and Society in Ancient
    Greece and China* / 1

2   The Social Origins of Mind
    Economics, Social Practices, and Thought / 29

3   Living Together vs. Going It Alone
    *Social Life and Sense of Self in the Modern
    East and West* / 47

4   "Eyes in Back of Your Head" or "Keep Your
    Eye on the Ball"?
    Envisioning the World / 79

5   "The Bad Seed" or "The Other Boys Made
    Him Do It"?
    *Causal Attribution and Causal Modeling
    East and West* / 111

6   Is the World Made Up of Nouns or Verbs?
    *Categories and Rules vs. Relationships and
    Similarities* / 137

7 "Ce N'est Pas Logique" or "You've Got a
Point There"?
*Logic and the Law of Noncontradiction vs.
Dialectics and the Middle Way* / 165

8 And If the Nature of Thought Is Not
Everywhere the Same?
*Implications for Psychology, Philosophy,
Education, and Everyday Life* / 191

Epilogue The End of Psychology or the Clash of
Mentalities?
The Longevity of Differences / 219

Notes / 231
References / 241
Index / 253

ACKNOWLEDGMENTS

When I read the acknowledgments in other people's books, I often wonder if all the people mentioned could really have made a significant contribution to the outcome. Please be assured that all the people mentioned below had a significant impact on this book, and some of them made enormous contributions.

The book would never have been written were it not for the fact that I have been blessed in recent years with some of the most remarkable students I have encountered in my entire professional life. Many of the ideas, especially for the experimental work, are theirs. The students include Incheol Choi, Marion Davis, Trey Hedden, Li-jun Ji, Jan Leu, Takahiko Masuda, Michael Morris, Ara Norenzayan, Kaiping Peng, and Jeffrey Sanchez-Burks.

Many of the ideas in the book have been shaped by discussions with colleagues in fields ranging from philosophy to physics. These include Susan Andersen, Scott Atran, Patricia Cheng, Lawrence Hirschfeld, Philip Ivanhoe, Qicheng Jing, Gordon Kane, Shinobu Kitayama, Hazel Markus, Donald Munro, Denise Park, Lee Ross, Edward E. Smith, Stephen Stich, and Frank Yates. They have given breadth and depth to my understanding of the East and to the framing of the ideas in this book.

Most of the students, collaborators, and colleagues went the extra mile and read early versions of the manuscript. They were joined by Richard Cassidy, Oona Cha, Dov Cohen, Joe Henrich, Peter Katzenstein, Joel Kupperman, Darrin Lehman, David Liu, Avashai Margalit, Yuri Miyamoto, Randolph Nesse, Yu Niiya, and Paolo Sousa.

I thank Philip Rappaport of Free Press for making editorial changes that made the book much clearer and more agreeable to read, and for helping me to avoid errors, and Philip Metcalf for an excellent job of copyediting. I'm grateful to my agents, John Brockman and Katina Matson, for representing my work and for their commitment to bringing science to a broad public.

I have been extremely fortunate in grant support. Institutions that provided valuable help in the form of research funds and release time include the John Simon Guggenheim Foundation, the National Institute of Aging, the National Science Foundation, the Russell Sage Foundation, and especially the University of Michigan and its Institute for Social Research in the form of generous support for the Culture and Cognition Program. Mary Cullen and Nancy Exelby of the Research Center for Group Dynamics of ISR and Eric Lomazoff of the Russell Sage Foundation provided excellent support for research and writing activities. They took burdens off of me and made chores into pleasures. Laura Reynolds of ISR solved problems I didn't know existed and provided extraordinarily able, cheerful, and willing help at every juncture.

Finally, my wife, Susan, provided invaluable advice and editorial suggestions, and, far more important, together with my children, Sarah and Matthew, sustained a life that gave meaning to the work. The book is dedicated to my son Matthew, whose interest in the East is as old as mine and whose youth has allowed him to learn far more from it than I could hope to do.

INTRODUCTION

A few years back, a brilliant student from China began to work with me on questions of social psychology and reasoning. One day early in our acquaintance, he said, "You know, the difference between you and me is that I think the world is a circle, and you think it's a line." Unfazed by what must have been a startled expression on my face, he expounded on that theme. "The Chinese believe in constant change, but with things always moving back to some prior state. They pay attention to a wide range of events; they search for relationships between things; and they think you can't understand the part without understanding the whole. Westerners live in a simpler, more deterministic world; they focus on salient objects or people instead of the larger picture; and they think they can control events because they know the rules that govern the behavior of objects."

I was skeptical but intrigued. I had been a lifelong universalist concerning the nature of human thought. Marching in step with the long Western line, from the British empiricist philosophers such as Hume, Locke, and Mill to modern-day cognitive scientists, I believed that all human groups perceive and reason in the same way. The shared assumptions of this tradition can be summarized with a few principles.

- Everyone has the same basic cognitive processes. Maori herders, !Kung hunter-gatherers, and dot-com entrepreneurs all rely on the same tools for perception, memory, causal analysis, categorization, and inference.
- When people in one culture differ from those in another in their beliefs, it can't be because they have different cognitive processes, but because they are exposed to different aspects of the world, or because they have been taught different things.
- "Higher order" processes of reasoning rest on the formal rules of logic: for example, the prohibition against contradiction—a proposition can't be both true and false.
- Reasoning is separate from what is reasoned about. The same process can be used to think about utterly different things and a given thing can be reasoned about using any number of different procedures.

A dozen years before meeting my student I had coauthored with Lee Ross a book with a title that made my sympathies clear—*Human Inference*. Not Western inference (and certainly not American college student inference!), but *human* inference. The book characterized what I took to be the inferential rules that people everywhere use to understand the world, including some rules that I believed were flawed and capable of producing erroneous judgments.

On the other hand, shortly before I met my new Chinese student, I had just completed a series of studies examining whether people's reasoning could be improved by teaching them new rules for thinking. Given my assumptions about universality and hard wiring, I had initially assumed the work would show that it is difficult, if not impossible, to change the patterns of reasoning I had been studying—even with immersion in long courses of study in fields such as statistics and economics. But to my surprise, I found substantial training effects. For example, people who have taken a few statistics courses avoid lots of errors in daily life: They're more likely to see that the "sophomore slump" in baseball could be due to statistical regression to the mean rather than to some mystical curse, and more likely to realize that an interview should be regarded as a small sample of a person's behavior and, therefore, that a wise hiring decision should be based on the larger sample of information in the application folder. Economists, it turns out, think differently about all sorts of things than the rest of us do—from deciding whether to remain at a boring movie to reasoning about foreign policy. Moreover, I found it was possible to train people in brief sessions and change not only their thinking habits, but their actual behavior when we tested them surreptitiously outside the laboratory.

So I was willing to give the student—whose name is Kaiping Peng and who now teaches at the University of California at Berkeley—an attentive hearing. If it's possible to produce marked changes in the way adults think, it certainly seemed possible that indoctrination into distinctive habits of thought from birth could result in very large *cultural* differences in habits of thought.

I began reading comparative literature on the nature of thought by philosophers, historians, and anthropologists—both Eastern and Western—and found that Peng had been a faithful reporter. Whereas psychologists have assumed universality, many scholars in other fields believe that Westerners (primarily Europeans, Americans, and citizens of the British Commonwealth) and East Asians (principally the people of China, Korea, and Japan) have maintained very different systems of thought for thousands of years. Moreover, these scholars are in substantial agreement about the nature of these differences. For example, most who have addressed the question hold that European thought rests on the assumption that the behavior of objects—physical, animal, and human—can be understood in terms of straightforward rules. Westerners have a strong interest in categorization, which helps them to know what rules to apply to the objects in question, and formal logic plays a role in problem solving. East Asians, in contrast, attend to objects in their broad context. The world seems more complex to Asians than to Westerners, and understanding events always requires consideration of a host of factors that operate in relation to one another in no simple, deterministic way. Formal logic plays little role in problem solving. In fact, the person who is too concerned with logic may be considered immature.

As a psychologist, I found these assertions to be revolutionary in their implications. If the scholars in the humanities and other social sciences were right, then the cognitive scientists were wrong: Human cognition is not everywhere the same. Without putting it in so many words, the humanities and social science scholars were making extremely important claims about the nature of thought. First, that members of different cultures differ in their "metaphysics," or fundamental beliefs about the nature of the world. Second, that the characteristic thought processes of different groups differ greatly. Third, that the thought processes are of a piece with beliefs about the nature of the world: People use the cognitive tools that seem to make sense—given the sense they make of the world.

Just as remarkably, the social structures and sense of self that are characteristic of Easterners and Westerners seem to fit hand in glove with their respective belief systems and cognitive processes. The collective or interdependent nature of Asian society is consistent with Asians' broad, contextual view of the world and their belief that events are highly complex and determined by many factors. The individualistic or independent nature of Western society seems consistent with the Western focus on particular objects in isolation from their context and with Westerners' belief that they can know the rules governing objects and therefore can control the objects' behavior.

If people really do differ profoundly in their systems of thought—their worldviews and cognitive processes— then differences in people's attitudes and beliefs, and even their values and preferences, might not be a matter merely of different inputs and teachings, but rather an inevitable consequence of using different tools to understand the world. And if that's true, then efforts to improve international understanding may be less likely to pay off than one might hope.

My student's chance comment, together with my interest in cultural psychology and the resulting reading program he had encouraged, launched me on a new course of research. I began a series of comparative studies, working with students at the University of Michigan and eventually with colleagues at Beijing University, Kyoto University, Seoul National University, and the Chinese Institute of Psychology. The research shows that there are indeed dramatic differences in the nature of Asian and European thought processes. The evidence lends support to the claims of nonpsychologist scholars and extends those claims to many surprising new mental phenomena. In addition, surveys and observational research document differences in social practices that dovetail with the differences in habits of thought. The new research has provided us, as prior evidence could not, with enough information so that we can build a theory about the nature of these differences, including how they might have come about, what their implications are for perceiving and reasoning in everyday life, and how they affect relations between people from different cultures.

The research allows us to answer many questions about social relations and thought that have long puzzled educators, historians, psychologists, and philosophers of science. Neither common stereotypical views about East-West differences nor the more sophisticated views of scholars can answer these questions or deal with the new findings. The puzzles and new observations range across many different domains. For example:

Science and Mathematics Why would the ancient
   Chinese have excelled at algebra and arithmetic
   but not geometry, which was the forte of the
   Greeks? Why do modern Asians excel at math
   and science but produce less in the way of revolutionary science than Westerners?

Attention and Perception Why are East Asians better
   able to see relationships among events than Westerners are? Why do East Asians find it relatively
   difficult to disentangle an object from its sur-
   roundings?

Causal Inference Why are Westerners so likely to
   overlook the influence of context on the behavior
   of objects and even of people? Why are Easterners more susceptible to the "hindsight bias," which
   allows them to believe that they "knew it all
   along"?

Organization of Knowledge Why do Western infants
   learn nouns at a much more rapid rate than verbs,
   whereas Eastern infants learn verbs at a more
   rapid rate than nouns? Why do East Asians group
   objects and events based on how they relate to
   one another, whereas Westerners are more likely
   to rely on categories?

Reasoning Why are Westerners more likely to apply
   formal logic when reasoning about everyday
   events, and why does their insistence on logic
   sometimes cause them to make errors? Why are
   Easterners so willing to entertain apparently contradictory propositions and how can this sometimes be helpful in getting at the truth?

Where to look for the causes of such vastly different systems of thought? Do they lie in biology? Language? Economics? Social systems? What keeps them going today? Social practices? Education? Inertia? And where are we headed with the differences? Will they still be here fifty or five hundred years from now?

My research has led me to the conviction that two utterly different approaches to the world have maintained themselves for thousands of years. These approaches include profoundly different social relations, views about the nature of the world, and characteristic thought processes. Each of these orientations—the Western and the Eastern—is a self-reinforcing, homeostatic system. The social practices promote the worldviews; the worldviews dictate the appropriate thought processes; and the thought processes both justify the worldviews and support the social practices. Understanding these homeostatic systems has implications for grasping the fundamental nature of the mind, for beliefs about how we ought ideally to reason, and for appropriate educational strategies for different peoples.

Perhaps most important of all, the book has implications for how East and West can get along better through mutual understanding of mental differences. Many people in Eastern countries believe with some justice that the past five hundred years of Western military, political, and economic dominance have made the West intellectually and morally arrogant. This book will have achieved its purpose for Western readers if it causes them to consider the possibility that another valid approach to thinking about the world exists and that it can serve as a mirror with which to examine and critique their own beliefs and habits of mind. The book will have served its purpose for Asian readers if it encourages them to consider the complementary possibility—though the need is perhaps less urgent for them because most Eastern intellectuals are already familiar to a considerable degree with Western ways of thinking.

To establish the contention that very different systems of perception and thought exist—and have existed for thousands of years—I draw on historical and philosophical evidence, as well as modern social science research, including ethnographies, surveys, and laboratory research. In chapter 1, Aristotle and Confucius are presented as examples of two different systems of thought. Undoubtedly those philosophers also served to entrench habits of thought that were already characteristic of their societies, but chapters 2 and 3 are intended to show that the social-practice differences found in modern societies would tend to sustain or even to create those different patterns even if they had not been present in ancient times. The heart of the book is contained in chapters 4 to 7. They present evidence that fundamental beliefs about the nature of the world, as well as the ways of perceiving it and reasoning about it, differ dramatically among modern peoples. The evidence is based in good part on laboratory research that I have conducted with students and colleagues using a variety of tests to examine how people perceive, remember, and think. Chapter 8 spells out some of the implications for psychology, philosophy, and society of the deep differences in systems of thought we have discovered. The epilogue speculates about where we are headed—toward convergence or toward continued or even intensified separation.

To set the stage a bit for the research that follows: When I speak of East Asia I mean China and the countries that were heavily influenced by its culture, most notably Japan and Korea. (I will sometimes abbreviate "East Asian" to "Easterner" and sometimes to "Asian.") When I speak of Westerners I mean people of European culture. When I speak of European Americans I mean blacks and whites and Hispanics—anyone but people of Asian descent. This somewhat odd usage can be justified by the fact that everyone born and raised in America is exposed to similar, though of course not by any means identical, cultural influences. This is true of Asian Americans too, obviously, but in some of the research discussed they are examined as a separate group because we would expect them to be more similar to Asians than we would expect other Americans to be—and in fact this is what we find.

Finally, I wish to apologize in advance to those people who will be upset to see billions of people labeled with the single term "East Asian" and treated as if they are identical. I do not mean to suggest that they are even close to being identical. The cultures and subcultures of the East differ as dramatically from one another as do those of the West. But the broad-brush term "East Asian" can be justified. In a host of social and political ways the cultures in that region are, in some general respects, similar to one another and different from Western countries. This will not satisfy some people who are highly knowledgeable about the East, but I ask them to bear with me. Some generalizations are justified despite the myriad differences. An analogy can be drawn to the study of language groups. Indo-European languages differ from one another in countless ways, and East Asian languages differ at least as much. Nevertheless, generalizations about the differences between Indo-European languages and East Asian languages taken as a group are possible and meaningful. And, as will be seen, some of those high-level generalizations are remarkably similar to some of the differences in perceptual and thought processes examined in this book.

CHAPTER 1

THE SYLLOGISM AND
THE TAO

Panoramic view of theat

More than a billion people in the world today claim intellectual inheritance from ancient Greece. More than two billion are the heirs of ancient Chinese traditions of thought. The philosophies and achievements of the Greeks and Chinese of 2,500 years ago were remarkably different, as were the social structures and conceptions of themselves. And, as I hope to show in this chapter, the intellectual aspects of each society make sense in light of their social characteristics.

THE ANCIENT GREEKS AND AGENCY

There is an ancient theater at Epidaurus in Greece that holds fourteen thousand people. Built into a hillside, the theater has a spectacular view of mountains and pine trees. Its acoustics are such that it is possible to hear a piece of paper being crumpled on the stage from any location in the theater. Greeks of the classical period, from the sixth to the third century B.C., traveled for long periods under difficult conditions to attend plays and poetry readings at Epidaurus from dawn till dusk for several days in a row.

To us today, people's love of the theater and their willingness to endure some hardship to indulge it may not seem terribly odd. But among the great civilizations of the day, including Persia, India, and the Middle East, as well as China, it is possible to imagine only the Greeks feeling free enough, being confident enough in their ability to control their own lives, to go on a long journey for the sole purpose of aesthetic enjoyment. The Greeks' contemporaries lived in more or less autocratic societies in which the king's will was law and to defy it was to court death. It would not have been in a ruler's interest to allow his subjects to wander about the countryside even if his subjects' ties to the land and the routines of agriculture had allowed them to imagine going on a long journey for purposes of recreation.

Equally astonishing, even to us today, is that the entire Greek nation laid down its tools—including its arms if city-states were at war with one another—to participate in the Olympics as athletes or audience.

The Greeks, more than any other ancient peoples, and in fact more than most people on the planet today, had a remarkable sense of personal *agency*—the sense that they were in charge of their own lives and free to act as they chose. One definition of happiness for the Greeks was that it consisted of being able to exercise their powers in pursuit of excellence in a life free from constraints.

A strong sense of individual identity accompanied the Greek sense of personal agency. Whether it is the Greeks or the Hebrews who invented individualism is a matter of some controversy, but there is no doubt that the Greeks viewed themselves as unique individuals, with distinctive attributes and goals. This would have been true at least by the time of Homer in the eighth or ninth century B.C. Both gods and humans in the *Odyssey* and the *Iliad* have personalities that are fully formed and individuated. Moreover, the differences among individuals were of substantial interest to Greek philosophers.

The Greek sense of agency fueled a tradition of debate. Homer makes it clear that a man is defined almost as much by his ability to debate as by his prowess as a warrior. A commoner could challenge even a king and not only live to tell the tale, but occasionally sway an audience to his side. Debates occurred in the marketplace, the political assembly, and even in military settings. Uniquely among ancient civilizations, great matters of state, as well as the most ordinary questions, were often decided by public, rhetorical combat rather than by authoritarian fiat. Tyrannies were not common in Greece and, when they arose, were frequently replaced by oligarchies or, beginning in the fifth century B.C., by democracies. The constitutions of some cities had mechanisms to prevent officials from becoming tyrants. For example, the city of Drerus on Crete prohibited a man from holding the office of *kosmos* (magistrate) until ten years had gone by since the last time he held the office.

As striking as the Greeks' freedom and individuality is their sense of curiosity about the world. Aristotle thought that curiosity was the uniquely defining property of human beings. St. Luke said of the Athenians of a later era: "They spend their time in nothing else but to tell or to hear some new thing." The Greeks, far more than their contemporaries, speculated about the nature of the world they found themselves in and created models of it. They constructed these models by categorizing objects and events and generating rules about them that were sufficiently precise for systematic description and explanation. This characterized their advances in—some have said invention of—the fields of physics, astronomy, axiomatic geometry, formal logic, rational philosophy, natural history, and ethnography. (The word "ethnocentric" is of Greek origin. The term resulted from the Greeks' recognition that their belief that their way of life was superior to that of the Persians might be based on mere prejudice. They decided it was not.)

Whereas many great contemporary civilizations, as well as the earlier Mesopotamian and Egyptian and the later Mayan civilizations, made systematic observations in all scientific domains, only the Greeks attempted to explain their observations in terms of underlying principles. Exploring these principles was a source of pleasure for the Greeks. Our word "school" comes from the Greek *scholē*, meaning "leisure." Leisure meant for the Greeks, among other things, the freedom to pursue knowledge. The merchants of Athens were happy to send their sons to school so that they could indulge their curiosity.

The Ancient Chinese and Harmony

While a special occasion for the ancient Greek might mean attendance at plays and poetry readings, a special occasion for the Chinese of the same period would be an opportunity to visit with friends and family. There was a practice called *chuan men*, literally "make doors a chain." Visits, which were intended to show respect for the hosts, were especially common during the major holidays. Those who were visited early were perceived as more important than those who were visited later.

The Chinese counterpart to Greek agency was *harmony*. Every Chinese was first and foremost a member of a collective, or rather of several collectives—the clan, the village, and especially the family. The individual was not, as for the Greeks, an encapsulated unit who maintained a unique identity across social settings. Instead, as philosopher Henry Rosemont has written: " . . . For the early Confucians, there can be no me in isolation, to be considered abstractly: I am the totality of roles I live in relation to specific others . . . Taken collectively, they weave, for each of us, a unique pattern of personal identity, such that if some of my roles change, the others will of necessity change also, literally making me a different person."

The Chinese were concerned less with issues of control of others or the environment than with self-control, so as to minimize friction with others in the family and village and to make it easier to obey the requirements of the state, administered by magistrates. The ideal of happiness was not, as for the Greeks, a life allowing the free exercise of distinctive talents, but the satisfactions of a plain country life shared within a harmonious social network. Whereas Greek vases and wine goblets show pictures of battles, athletic contests, and bacchanalian parties, ancient Chinese scrolls and porcelains depict scenes of family activities and rural pleasures.

The Chinese would not have felt themselves to be the helpless pawns of superiors and family members. On the contrary, there would have been a sense of *collective agency*. The chief moral system of China—Confucianism —was essentially an elaboration of the obligations that obtained between emperor and subject, parent and child, husband and wife, older brother and younger brother, and between friend and friend. Chinese society made the individual feel very much a part of a large, complex, and generally benign social organism where clear mutual obligations served as a guide to ethical conduct. Carrying out prescribed roles—in an organized, hierarchical system— was the essence of Chinese daily life. There was no counterpart to the Greek sense of personal liberty. Individual rights in China were one's "share" of the rights of the community as a whole, not a license to do as one pleased.

Within the social group, any form of confrontation, such as debate, was discouraged. Though there was a time, called the period of the "hundred schools" of 600 to 200 B.C., during which polite debate occurred, at least among philosophers, anything resembling public disagreement was discouraged. As the British philosopher of science Geoffrey Lloyd has written, "In philosophy, in medicine, and elsewhere there is criticism of other points of view . . . [but] the Chinese generally conceded far more readily than did the Greeks, that other opinions had something to be said for them ..."

Their monophonic music reflected the Chinese concern with unity. Singers would all sing the same melody and musical instruments played the same notes at the same time. Not surprisingly, it was the Greeks who invented polyphonic music, where different instruments, and different voices, take different parts.

Chinese social harmony should not be confused with conformity. On the contrary, Confucius praised the desire of the gentleman to harmonize and distinguished it from the petty person's need for conformity. The *Zuozhuan*, a classic Confucian text, makes the distinction in a metaphor about cooking. A good cook blends the flavors and creates something harmonious and delicious. No flavor is completely submerged, and the savory taste is due to the blended but distinctive contributions of each flavor.

The Chinese approach to understanding the natural world was as different from that of the Greeks as their understanding of themselves. Early in their study of the heavens, the Chinese believed that cosmic events such as comets and eclipses could predict important occurrences on earth, such as the birth of conquerors. But when they discovered the regularities in these events, so far from building models of them, they lost interest in them.

The lack of wonder among the Chinese is especially remarkable in light of the fact that Chinese civilization far outdistanced Greek civilization technologically. The Chinese have been credited with the original or independent invention of irrigation systems, ink, porcelain, the magnetic compass, stirrups, the wheelbarrow, deep drilling, the Pascal triangle, pound locks on canals, fore-and-aft sailing, watertight compartments, the sternpost rudder, the pad-dle-wheel boat, quantitative cartography, immunization techniques, astronomical observations of novae, seismographs, and acoustics. Many of these technological achievements were in place at a time when Greece had virtually none.

But, as philosopher Hajime Nakamura notes, the Chinese advances reflected a genius for practicality, not a penchant for scientific theory and investigation. And as philosopher and sinologist Donald Munro has written, "In Confucianism there was no thought of *knowing* that did not entail some consequence for action."

ESSENCE OR EVANESCENCE?

PHILOSOPHY IN GREECE AND CHINA

The philosophies of Greece and China reflected their distinctive social practices. The Greeks were concerned with understanding the fundamental nature of the world, though in ways that were different in different eras. The philosophers of Ionia (including western Turkey, Sicily, and southern Italy) of the sixth century B.C. were thoroughly empirical in orientation, building their theories on a base of sense observation. But the fifth century saw a move toward abstraction and distrust of the senses. Plato thought that ideas—the *forms*—had a genuine reality and that the world could be understood through logical

approaches to their meaning, without reference to the
world of the senses. If the senses seemed to contradict
conclusions reached from first principles and logic, it was
the senses that had to be ignored.

Though Aristotle did not grant reality to the forms, he
thought of attributes as having a reality distinct from their
concrete embodiments in objects. For him it was meaningful to speak not just of a solid object, but of attributes in
the abstract—solidity, whiteness, etc.—and to have theories about these abstractions. The central, basic, sine qua
non properties of an object constituted its "essence," which
was unchanging by definition, since if the essence of an
object changed it was no longer the object but something
else. The properties of an object that could change without changing the object's essence were "accidental" properties. For example, the author is sadly lacking in musical
talent, but if he suddenly were to have musical talent, you
would still think he was the same person. Musical talent,
then, is an accidental property, and change in it does not
constitute change in the person's essence. Greek philosophy thus differed greatly from Chinese in that it was
deeply concerned with the question of which properties
made an object what it was, and which were alterable
without changing the nature of the object.

The Greek language itself encouraged a focus on
attributes and on turning attributes into abstractions. As in
other Indo-European languages, every adjective can be
granted noun status by adding the English equivalent of
"ness" as a suffix: "white" becomes "whiteness"; "kind"
becomes "kindness." A routine habit of Greek philosophers
was to analyze the attributes of an object—person, place,

thing, or animal—and categorize the object on the basis of its abstracted attributes. They would then attempt to understand the object's nature, and the cause of its actions, on the basis of rules governing the categories. So the attributes of a comet would be noted and the object would then be categorized at various levels of abstraction—*this* comet, *a* comet, a heavenly body, a moving object. Rules at various levels of abstraction would be generated as hypotheses and the behavior of the comet explained in terms of rules that seemed to work at a given level of abstraction.

But still more basic to Greek philosophy is its background scheme, which regarded the object *in isolation* as the proper focus of attention and analysis. Most Greeks regarded matter as particulate and separate—formed into discrete objects—just as humans were seen as separate from one another and construed as distinct wholes. Once the object is taken as the starting point, then many things follow automatically: The attributes of the object are salient; the attributes become the basis of categorization of the object; the categories become the basis of rule construction; and events are then understood as the result of objects behaving in accordance with rules. By "objects" I mean both nonhuman and human objects, but in fact the nature of the physical world was of great concern to Greek philosophers. Human relations and ethical conduct were important to the Greeks but did not have the consuming interest that they did for the Chinese.

A peculiar but important aspect of Greek philosophy is the notion that the world is fundamentally static and unchanging. To be sure, the sixth-century philosopher

Heraclitus and other early philosophers were concerned with change. ("A man never steps in the same river twice because the man is different and the river is different.") But by the fifth century, change was out and stability was in. Parmenides "proved," in a few easy steps, that change was impossible: To say of a thing that it does not exist is a contradiction. Nonbeing is self-contradictory and so nonbeing can't exist. If nonbeing can't exist, then nothing can change because, if thing 1 were to change to thing 2, then thing 1 would not be! Parmenides created an option for Greek philosophers: They could trust either logic or their senses. From Plato on, they often went with logic.

Zeno, the pupil of Parmenides, established in a similar way that motion was impossible. He did this in two demonstrations. One is his famous demonstration with the arrow. In order for an arrow to reach a target, it first has to go halfway toward the target, then halfway between that and the target, and then halfway between *that* and the target, etc. But of course half of a half of a half . . . still leaves the arrow short of the target. Ergo, visual evidence to the contrary notwithstanding, movement can't occur. The other "proof" was even simpler. Either a thing is in its place or it is not. If it is in its place, then it cannot move. It is impossible for a thing not to be in its place; therefore nothing moves. As communications theorist Robert Logan has written, the Greeks "became slaves to the linear, either-or orientation of their logic."

Not all Greek philosophers were logic-choppers out to prove change impossible, but there is a static quality even to the reasoning of Aristotle. He believed, for example, that all celestial bodies were immutable, perfect spheres and though motion occurs and events happen, the essences of things do not change. Moreover, Aristotle's physics is highly linear. Changes in rate of motion, let alone cyclical motion, play little role in Aristotle's physics. (It is partly for this reason that Aristotle's physics was so remarkably misguided. Gordon Kane, a physicist friend of mine, has identified a large number of physical propositions in Aristotle's writings. He maintains that the great majority of them are wrong. This is especially puzzling because Aristotle's Ionian predecessors got many of them right.)

The Chinese orientation toward life was shaped by the blending of three different philosophies: Taoism, Confucianism, and, much later, Buddhism. Each philosophy emphasized harmony and largely discouraged abstract speculation.

There is an ancient Chinese story, still known to most East Asians today, about an old farmer whose only horse ran away. Knowing that the horse was the mainstay of his livelihood, his neighbors came to commiserate with him. "Who knows what's bad or good?" said the old man, refusing their sympathy. And indeed, a few days later his horse returned, bringing with it a wild horse. The old man's friends came to congratulate him. Rejecting their congratulations, the old man said, "Who knows what's bad or good?" And, as it happened, a few days later when the old man's son was attempting to ride the wild horse, he was thrown from it and his leg was broken. The friends came to express their sadness about the son's misfortune. "Who knows what's bad or good?" said the old man. A few

weeks passed, and the army came to the village to con-
script all the able-bodied men to fight a war against the
neighboring province, but the old man's son was not fit to
serve and was spared.

The story, which goes on as long as the patience of the
audience permits, expresses a fundamental of the Eastern
stance toward life. The world is constantly changing and is
full of contradictions. To understand and appreciate one
state of affairs requires the existence of its opposite; what
seems to be true now may be the opposite of what it
seems to be (cf. Communist-era Premier Chou En-lai's
response when asked whether he thought the consequences of the French Revolution had been beneficial: "It's
too early to tell").

The sign of the Tao.

Yin (the feminine and dark and passive) alternates
with yang (the masculine and light and active). Indeed yin
and yang only exist because of each other, and when the
world is in a yin state, this is a sure sign that it is about to
be in a yang state. The sign of the Tao, which means "the

Way" to exist with nature and with one's fellow humans, consists of two forces in the form of a white and a black swirl. But the black swirl contains a white dot and the white swirl contains a black dot. And "the truest yang is the yang that is in the yin." The principle of yin-yang is the expression of the relationship that exists between opposing but interpenetrating forces that may complete one another, make each comprehensible, or create the conditions for altering one into the other.

From the *I Ching*: " . . . For misery, happiness is leaning against it; for happiness, misery is hiding in it. Who knows whether it is misery or happiness? There is no certainty. The righteous suddenly becomes the vicious, the good suddenly becomes the bad" (*I Ching*, xxx).

From the *Tao Te Ching*: "The heavy is the root of the light . . . The unmoved is the source of all movement" (Chapter 26).

Returning—moving in endless cycles—is the basic pattern of movement of the Tao.

> *To shrink something*
> *You need to expand it first*
> *To weaken something*
> *You need to strengthen it first*
> *To abolish something*
> *You need to flourish it first*
> *To take something*
> *You need to give it first* (Tao Te Ching, *Chapter 36)*

Aside from Taoism's teachings about opposition, contradiction, change, and cycles, it stood for a deep appreciation of nature, the rural life, and simplicity. It was the religion of wonder, magic, and fancy, and it gave meaning to the universe through its account of the links between nature and human affairs.

Taoism is the source of much of the philosophy behind the healing arts of China. Physiology was explained on a symbolic level by the yin-yang principle and by the Five Elements (earth, fire, water, metal, and wood), which also provided the explanations behind magic, incantations, and aphrodisiacs. The ubiquitous word was *ch'i*, meaning variously "breath," "air," or "spirit."

Confucius, who lived from 551 to 479 B.C., was less a religious leader than an ethical philosopher. His concern was with the proper relations among people, which in his system were hierarchical and strictly spelled out. Each member of each of the important relationship pairs (husband-wife, etc.) had clear obligations toward the other.

Confucianism has been called the religion of common sense. Its adherents are urged to uphold the Doctrine of the Golden Mean—to be excessive in nothing and to assume that between two propositions, and between two contending individuals, there is truth on both sides. But in reality, Confucianism, like Taoism, is less concerned with finding the truth than with finding the Tao—the Way—to live in the world.

Confucianism stresses economic well-being and education. The individual works not for self-benefits but for the entire family. Indeed, the concept of self-advancement, as opposed to family advancement, is foreign to cultures that are steeped in the Confucian orientation. A promising young man was expected to study for the government examinations with the hope of becoming a magistrate. If he did, his whole family benefited economically from his position. Unlike most of the world until very modern times, there was substantial social and economic mobility in China. Everyone who lived long enough would see families rise far higher than their origins and others sink far lower. Perhaps partly for this reason, Confucians have always believed, far more than the intellectual descendants of Aristotle, in the malleability of human nature.

Confucianism blended smoothly with Taoism. In particular, the deep appreciation of the contradictions and changes in human life, and the need to see things whole, that are integral to the notion of a yin-yang universe are also part of Confucian philosophy. But the dominant themes of nature and the rural life are much more associated with Taoism than with Confucianism, and the importance of the family and educational and economic advancement are more integral to Confucianism. These thematic differences are reflected in paintings on porcelains and scrolls. Characteristic Tao-inspired themes would include a picture of a fisherman, a woodcutter, or a lone individual sitting under trees. Confucian-inspired themes would center on the family, with pictures of many people of different ages engaging in shared activities. Different individuals in ancient China, and for that matter in contemporary China, would likely emphasize one of the orientations more than the other. This might depend in part on station in life. There is an adage holding that every Chinese is a Confucianist when he is successful and a Taoist when he is a failure.

Buddhism came to China from India hundreds of years

after the classical period we are discussing. The Chinese readily absorbed congenial aspects of Buddhism, including what had been missing in Chinese philosophy, notably an epistemology, or theory of knowledge. All three orientations shared concerns about harmony, holism, and the mutual influence of everything on almost everything else. These orientations help explain why Chinese philosophy not only lacked a conception of individual rights but, it sometimes seems (at least after Buddhism began to exert an influence), an acknowledgment of individual minds. A twelfth-century neo-Confucian wrote, "The universe is my mind and my mind is the universe. Sages appeared tens of thousands of generations ago. They shared this mind; they shared this principle. Sages will appear tens of thousands of generations to come. They will share this mind; they will share this principle."

The holism common to the three orientations suggested that every event is related to every other event. A key idea is the notion of resonance. If you pluck a string on an instrument, you produce a resonance in another string. Man, heaven, and earth create resonances in each other. If the emperor does something wrong, it throws the universe out of kilter.

The concern with abstraction characteristic of ancient Greek philosophy has no counterpart in Chinese philosophy. Chinese philosophers quite explicitly favored the most concrete sense impressions in understanding the world. In fact, the Chinese language itself is remarkably concrete. There is no word for "size," for example. If you want to fit someone for shoes, you ask them for the "big-small" of their feet. There is no suffix equivalent to "ness" in Chinese. So there is no "whiteness"—only the white of the swan and the white of the snow. The Chinese are disinclined to use precisely defined terms or categories in any arena, but instead use expressive, metaphoric language.

In Chinese literary criticism there are different methods of writing called "the method of watching a fire across the river" (detachment of style), "the method of dragonflies skimming across the water surface" (lightness of touch), "the method of painting a dragon and dotting its eyes" (bringing out the salient points).

For the Chinese, the background scheme for the nature of the world was that it was a mass of substances rather than a collection of discrete objects. Looking at a piece of wood, the Chinese philosopher saw a seamless whole composed of a single substance, or perhaps of interpenetrating substances of several kinds. The Greek philosopher would have seen an object composed of particles. Whether the world was composed of atoms or of continuous substances was debated in Greece, but the issue never arose in China. It was continuous substances, period. Philosopher of science Joseph Needham has observed: "Their universe was a continuous medium or matrix within which interactions of things took place, not by the clash of atoms, but by radiating influences."

So the philosophies of China and Greece were as different as their respective social life and self-conceptions.

And the philosophical differences are reflective of the social ones, in several respects.

Greeks were independent and engaged in verbal contention and debate in an effort to discover what people took to be the truth. They thought of themselves as individuals with distinctive properties, as units separate from others within the society, and in control of their own destinies. Similarly, Greek philosophy started from the individual object—the person, the atom, the house—as the unit of analysis and it dealt with properties of the object. The world was in principle simple and knowable: All one had to do was to understand what an object's distinctive attributes were so as to identify its relevant categories and then apply the pertinent rule to the categories.

Chinese social life was interdependent and it was not liberty but harmony that was the watchword—the harmony of humans and nature for the Taoists and the harmony of humans with other humans for the Confucians. Similarly, the Way, and not the discovery of truth, was the goal of philosophy. Thought that gave no guidance to action was fruitless. The world was complicated, events were interrelated, and objects (and people) were connected "not as pieces of pie, but as ropes in a net." The Chinese philosopher would see a family with interrelated members where the Greek saw a collection of persons with attributes that were independent of any connections with others. Complexity and interrelation meant for the Chinese that an attempt to understand the object without appreciation of its context was doomed. Under the best of circumstances, control of outcomes was difficult.

Science and mathematics, as we'll see next, were fully consistent with both social behavior and philosophical outlook.

CONTRADICTION OR CONNECTION?

SCIENCE AND MATHEMATICS IN GREECE AND CHINA

The greatest of all Greek scientific discoveries was the discovery—or rather, as philosopher Geoffrey Lloyd put it, the invention—of nature itself. The Greeks defined nature as the universe minus human beings and their culture. Although this seems to us to be the most obvious sort of distinction, no other civilization came upon it. A plausible account of how the Greeks happened to invent nature is that they came to make a distinction between the external, objective world and the internal, subjective one. And this distinction came about because the Greeks, unlike everyone else, had a clear understanding of subjectivity arising from the tradition of debate. It makes no sense for you to try to persuade me of something unless you believe that there is a reality out there that you apprehend better than I do. You may be able to coerce me into doing what you want and even into saying that I believe what you do. But you will not persuade me until I believe that your subjective interpretation of some state of affairs is superior to mine.

So, in effect, objectivity arose from subjectivity—the recognition that two minds could have different representations of the world and that the world has an existence independent of either representation. This recognition was probably aided for the Greeks because, due to their position as a trading center, they regularly encountered people with utterly different notions about the world. In contrast, Chinese culture was unified early on and it would have been relatively rare to encounter people having radically different metaphysical and religious views.

The Greeks' discovery of nature made possible the invention of science. China's failure to develop science can be attributed in part to lack of curiosity, but the absence of a concept of nature would have blocked the development of science in any case. As philosopher Yu-lan Fung observes, "Why" questions are hard to ask if there is no clear recognition that there are mental concepts that somehow correspond to aspects of nature, but which are not identical to them.

The Greeks' focus on the salient object and its attributes led to their failure to understand the fundamental nature of causality. Aristotle explained that a stone falling through the air is due to the stone having the property of "gravity." But of course a piece of wood tossed into water floats instead of sinking. This phenomenon Aristotle explained as being due to the wood having the property of "levity"! In both cases the focus is exclusively on the object, with no attention paid to the possibility that some force outside the object might be relevant. But the Chinese saw the world as consisting of continuously interacting substances, so their attempts to understand it caused them to be oriented toward the complexities of the entire "field," that is, the context or environment as a whole. The notion that events always occur in a field of forces would have been completely intuitive to the Chinese. The Chinese therefore had a kind of recognition of the principle of "action at a distance" two thousand years before Galileo articulated it. They had knowledge of magnetism and acoustic resonance, for example, and believed it was the movement of the moon that caused the tides, a fact that eluded even Galileo.

In the desert of western China are buried bodies of tall, red-haired people, astonishingly well preserved, of Caucasian appearance. They found their way to that part of the world some thousands of years ago. Aside from the way they look, they are different from the peoples who lived in the area in another interesting respect. Many of them show clear signs of having been operated on surgically. In all of Chinese history, surgery has been a great rarity.

The reluctance of the Chinese to perform surgery is completely understandable in light of their views about harmony and relationships. Health was dependent on the balance of forces in the body and the relationships between its parts. And there were, and are for many East Asians today, relationships between every part of the body and almost every other part. To get a feel for this vast web of interconnections, look at a modern acupuncturist's view of the relations between the surface of the ear and the epidermis and skeleton. An equally complex network describes the relations between the ear and each of the internal organs. The notion that the removal of a malfunctioning or diseased part of the body could be beneficial, without attending to its relations to other parts of the body, would have been too simple-minded for the Chinese to contemplate. In contrast, surgery has been practiced in many different Western societies for thousands of years.

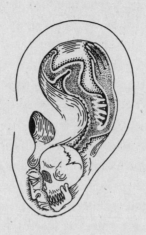

Representation of epidermis and skeleton on the surface
of the ear for purposes of acupuncture.

The Chinese tendency to focus on relationships in a complex, interconnected field is exemplified by the practice of feng shui, still continued in the East. When someone wishes to build a building, it is essential to call in a feng shui master. This person takes account of a seemingly limitless number of factors such as altitude, prevailing wind, orientation toward the compass, proximity to various bodies of water, and gives advice on where to locate the structure. This practice has had no real counterpart in the West, but the most modern skyscraper in Hong Kong will have had its feng shui workup before being built.

The Chinese conviction about the fundamental relatedness of all things made it obvious to them that objects are altered by context. Thus any attempt to categorize objects with precision would not have seemed to be of much help in comprehending events. The world was simply too complex and interactive for categories and rules to be helpful for understanding objects or controlling them.

The Chinese were right about the importance of the field to an understanding of the behavior of the object and they were right about complexity, but their lack of interest in categories prevented them from discovering laws that really were capable of explaining classes of events. And for all that the Greeks tended to oversimplify and to be satisfied by pseudo-explanations involving nonexistent properties of objects, they correctly understood that it was necessary to categorize objects in order to be able to apply rules to them. Since rules are useful to the extent that they apply to the widest possible array of objects, there was a constant "upward press" to generalize to high levels of abstraction so that rules would be maximally applicable. This drive toward abstraction was sometimes—though not always—useful.

The Greek faith in categories had scientific payoffs, immediately as well as later, for their intellectual heirs. Only the Greeks made classifications of the natural world sufficiently rigorous to permit a move from the sorts of folk-biological schemes that other peoples constructed to a single classification system that ultimately could result in theories with real explanatory power.

A group of mathematicians associated with Pythagoras is said to have thrown a man overboard because it was discovered that he had revealed the scandal of irrational numbers, such as the square root of 2, which just goes on and on without a predictable pattern: 1.4142135. . . . Whether this story is apocryphal or not, it is certainly the case that most Greek mathematicians did not regard irrational numbers as real numbers at all. The Greeks lived in a world of discrete particles and the continuous and unending nature of irrational numbers was so implausible that mathematicians could not take them seriously.

On the other hand, the Greeks were probably pleased by how it was they came to know that the square root of 2 is irrational, namely via a proof from contradiction. One posits two whole numbers, n and m, such that the square root of $2 = n/m$ and shows that this leads to a contradiction.

The Greeks were focused on, you might even say obsessed by, the concept of contradiction. If one proposition was seen to be in a contradictory relation with another, then one of the propositions had to be rejected. The principle of noncontradiction lies at the base of propositional logic. The general explanation given for why the Greeks, rather than some other people, invented logic, is that a society in which debate plays a prominent role will begin to recognize which arguments are flawed by definition because their structure results in a contradiction. The basic rules of logic, including syllogisms, were worked out by Aristotle. He is said to have invented logic because he was annoyed at hearing bad arguments in the political assembly and in the agora! Notice that logical analysis is a kind of continuation of the Greek tendency to decontextualize. Logic is applied by stripping away the meaning of statements and leaving only their formal structure intact. This makes it easier to see whether an argument is valid or not. Of course, as modern East Asians are fond of pointing out, that sort of decontextualization is not without its dangers. Like the ancient Chinese, they strive to be reasonable, not rational. The injunction to avoid extremes can be as useful a principle as the requirement to avoid contradictions.

Chinese philosopher Mo-tzu made serious strides in the direction of logical thought in the fifth century B.C., but he never formalized his system and logic died an early death in China. Except for that brief interlude, the Chinese lacked not only logic, but even a principle of contradiction. India did have a strong logical tradition, but the Chinese translations of Indian texts were full of errors and misunderstandings. Although the Chinese made substantial advances in algebra and arithmetic, they made little progress in geometry because proofs rely on formal logic, especially the notion of contradiction. (Algebra did not become deductive until Descartes. Our educational system retains the memory trace of their separation by teaching algebra and geometry as separate subjects.)

The Greeks were deeply concerned with foundational arguments in mathematics. Other peoples had recipes; only the Greeks had derivations. On the other hand, Greek logic and foundational concern may have presented as many obstacles as opportunities. The Greeks never developed the concept of zero, which is required both for algebra and for an Arabic-style place number system. Zero was considered by the Greeks, but rejected on the grounds that it represented a contradiction. Zero equals nonbeing and nonbeing cannot be! An understanding of zero, as well as of infinity and infinitesimals, ultimately had to be imported from the East.

In place of logic, the Chinese developed a type of *dialecticism*. This is not quite the same as the Hegelian dialectic in which thesis is followed by antithesis, which is resolved by synthesis, and which is "aggressive" in the sense that the ultimate goal of reasoning is to resolve contradiction. The Chinese dialectic instead uses contradiction to understand relations among objects or events, to transcend or integrate apparent oppositions, or even to embrace clashing but instructive viewpoints. In the Chinese intellectual tradition there is no necessary incompatibility between the belief that A is the case and the belief that not-A is the case. On the contrary, in the spirit of the Tao or yin-yang principle, A can actually imply that not-A is also the case, or at any rate soon will be the case. Dialectical thought is in some ways the opposite of logical thought. It seeks not to decontextualize but to see things in their appropriate contexts: Events do not occur in isolation from other events, but are always embedded in a meaningful whole in which the elements are constantly changing and rearranging themselves. To think about an object or event in isolation and apply abstract rules to it is to invite extreme and mistaken conclusions. It is the Middle Way that is the goal of reasoning.

Why should the ancient Greeks and Chinese have differed so much in their habits of thought or, at any rate, why should this be true of the intelligentsia, who are the only ancient peoples whose mental life is known to us at all? And why should there be such "resonance" between the social forms and self-understandings on the one hand and the philosophical assumptions and scientific approaches on the other? Answers to these questions have implications for understanding the differences between Eastern and Western thought that exist today.

CHAPTER 2

THE SOCIAL ORIGINS
OF MIND

I once asked a Chinese philosopher why he thought the
East and the West had developed such different habits of
thought. "Because you had Aristotle and we had Confucius," he replied. He was joking—mostly. Although Aristotle and Confucius had enormous impact on the
intellectual, social, and political histories of the peoples
who followed, they were less the progenitors of their
respective cultures than the products. And they couldn't
have had the impact they did if they hadn't reflected the
societies they lived in. In fact, a kind of "proof" of this is
that Greece did have its philosophers, like Heraclitus, who
were more nearly Eastern in spirit than Western, and
China had its philosophers, like Mo-tzu, who shared many
of the concerns of Western philosophers. But despite
receiving a good deal of attention from contemporaries,

the maverick philosophies died on the vine, and it is the Aristotelian tradition that continues in the West and the Confucian that continues in the East.

Scholars who have addressed the question of why ancient China and Greece differed so much have come up with several plausible reasons.

Greece differed from all contemporary civilizations in the development of personal freedom, individuality, and objective thought. These qualities seem partly explainable by the political system that was unique to Greece, namely the city-state and its politics, especially the assembly, in which people had to persuade one another by dint of rational argument. The city-state was also important because it was possible for intellectual rebels to leave one location and go to another, thereby maintaining a condition of relatively free inquiry. Indeed, members of the intelligentsia who were personae non gratae in a given city-state would sometimes be sought out by other city-states for the prestige they would bring. Socrates' followers begged him to leave Athens and go somewhere else rather than allow the death sentence against him to be carried out. He would have been welcomed elsewhere and there would have been no stomach for pursuit of him by his fellow citizens.

Another factor sometimes invoked to explain Greece's uniqueness is that its maritime location made trading a lucrative occupation, which meant that there was a substantial mercantile class who could afford to have their sons educated. That the merchants would have wished to have their sons educated requires explanation in itself, of course, especially because, unlike in China, education was

not a route to power and wealth. The drive toward education was apparently the result of curiosity and a belief in
the value of knowledge for its own sake. The curiosity
characteristic of the Greeks may in turn be explained in
part by the location of the Greeks at a crossroads of the
world. They were constantly encountering novel and per-
plexing people, customs, and beliefs. For any Greek living
near the coasts (and that would have been the great
majority), encountering people representing other ethnici-
ties, religions, and polities would have been common.
Athens itself would have been rather like the bar in *Star
Wars*.

An obvious consequence of the different practices and
beliefs swirling around the Greeks would have been the
necessity of dealing with contradiction. They would have
been constantly confronting situations where one person
was asserting that A was the case and another was contending that not-A was the case. Contradiction coming
from the opinions of outsiders, as well as freely expressed
contradiction among insiders' views in the assembly and
the marketplace, might have forced the development of
cognitive procedures, including formal logic, to deal with
the dissonance.

In contrast, even today 95 percent of the Chinese population belongs to the same Han ethnic group. Nearly all
of the country's more than fifty minority ethnic groups are
in the western part of the country. A Chinese person living
in the rest of the country would rarely have encountered
anyone having significantly different beliefs or practices.
The ethnic homogeneity of China seems at least partly
explicable in terms of the centralized political control. In

addition, the face-to-face village life of China would have pressed in the direction of harmony and agreed-upon norms for behavior. Seeing little difference of opinion, and finding disagreement sanctioned from above or from peers where it did exist, the Chinese would have had little use for procedures to decide which of two propositions was correct. Instead, finding means to resolve disagreements would have been the goal. Hence, the push to find the Middle Way.

HOMEOSTATIC SOCIO-COGNITIVE SYSTEMS

At base, all of these explanations rest on one fact: The ecologies of ancient Greece and China were drastically different—in ways that led to different economic, political, and social arrangements. The left side of the illustration that follows shows an account of the differences between Greek and Chinese thought that makes sense to me. It is essentially a distillation of the views of many people who have tackled the question of the origin of mentalities. The right side of the illustration is the same account, but drawn by a Chinese American student, who told me she felt that a circular presentation made more sense than my linear one!

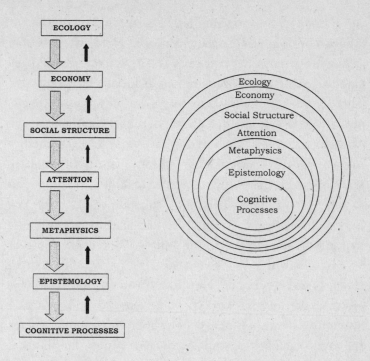

Schematic model of influences on cognitive processes.

The account is at base materialistic: That is, it attempts to explain cultural facts in terms of physical ones. This approach is currently out of fashion in some circles partly because it is assumed, mistakenly, that materialistic accounts are deterministic. But materialism need not imply inevitability—just that, other things equal, physical factors can influence to some degree economic factors and consequently cultural ones. The account is not at all materialistic in one sense: The critical factors influencing habits of mind are social and important social facts can be generated and sustained by forces that are not economic in nature.

Ecology → Economy and Social Structure The ecology of China, consisting as it does primarily of relatively fertile plains, low mountains, and navigable rivers, favored agriculture and made centralized control of society relatively easy. Agricultural peoples need to get along with one another—not necessarily to like one another (think of the stereotype of the crusty New England farmer)—but to live together in a reasonably harmonious fashion. This is particularly true for rice farming, characteristic of southern China and Japan, which requires people to cultivate the land in concert with one another. But it is also important wherever irrigation is required, as in the Yellow River Valley of north China, where the Shang dynasty (from the eighteenth to the eleventh century B.C.) and the Chou dynasty (from the eleventh century B.C. to 256 B.C.) were based. In addition to getting along with one's neighbors, irrigation systems require centralized control and ancient China, like all other ancient agricultural societies, was ruled by despots. Peasants had to get along with their neighbors and were ruled by village elders and a regional magistrate who was the representative of the king (and after the unification of China, of the emperor). The ordinary Chinese therefore lived in a complicated world of social constraints.

The ecology of Greece, on the other hand, consisting as it does mostly of mountains descending to the sea, favored hunting, herding, fishing, and trade (and—let's be frank—piracy). These are occupations that require relatively little cooperation with others. In fact, with the exception of trade, these economic activities do not strictly require living in the same stable community with other people. Settled agriculture came to Greece almost

two thousand years later than to China, and it quickly
became commercial, as opposed to merely subsistence, in
many areas. The soil and climate of Greece were congenial
to wine and olive oil production and, by the sixth century
B.C., many farmers were more nearly businessmen than
peasants. The Greeks were therefore able to act on their
own to a greater extent than were the Chinese. Not feeling it necessary to maintain harmony with their fellows at
any cost, the Greeks were in the habit of arguing with one
another in the marketplace and debating one another in
the political assembly.

*Social Structure and Social Practice → Attention and
Folk Metaphysics* The Chinese had to look outward toward
their peers and upward toward authorities in the conduct
of their economic, social, and political lives. Their relations
with others provided both the chief constraint in their
lives and the primary source of opportunities. The habit of
looking toward the social world could have carried over to
a tendency to look to the field in general; and the need to
attend to social relations could have extended to an inclination to attend to relations of all kinds. As social psychologists Hazel Markus and Shinobu Kitayama put it, "If one
perceives oneself as embedded within a larger context of
which one is an interdependent part, it is likely that other
objects or events will be perceived in a similar way." "Folk
metaphysics"—beliefs about the nature of the social and
physical world—would therefore both have been generated by one fact: the Chinese were attending closely to the
social world. The sense that the self was linked in a network of relationships and social obligations might have
made it natural to view the world in general as continuous

and composed of substances rather than discrete and consisting of distinct objects. Causality would be seen as being
located in the field or in the relation between the object
and the field. Attention to the field would encourage
recognition of complexity and change, as well as of contradiction among its many and varied elements.

But the Greeks had the luxury of attending to objects,
including other people and their own goals with respect to
them, without being overly constrained by their relations
with other people. A Greek could plan a harvest, arrange
for a relocation of his herd of sheep, or investigate
whether it would be profitable to sell some new commod-
ity, consulting little or not at all with others. This might
have made it natural for the Greeks to focus on the attributes of objects with a view toward categorizing them and
finding the rules that would allow prediction and control
of their behavior. Causality would be seen as due to properties of the object or as the result of one's own actions in
relation to the object. Such a view of causality could have
encouraged the Greek assumptions of stability and perma-
nence as well as an assumption that change in the object
was under their control.

So the folk metaphysics of the two societies could
have arisen directly from the targets of attention: the environment or field in the case of the Chinese and the object
in the case of the Greeks. The scientific metaphysics of
each society would have been just a reflection of the folk
views.

Folk Metaphysics → *Tacit Epistemology and Cognitive
Processes* Folk metaphysics can be expected to influence

tacit epistemology, or beliefs about how to get new knowledge. If the world is a place where relations among objects
and events are crucial in determining outcomes, then it
will seem important to be able to observe all the important elements in the field, to see relations among objects
and to see the relation between the parts and the whole.
Processes of attention, perception, and reasoning will
develop that focus on detecting the important events and
discerning the complex relationships among them. If, on
the other hand, the world is a place where the behavior of
objects is governed by rules and categories, then it should
seem crucial to be able to isolate the object from its context, to infer what categories the object is a member of,
and to infer how rules apply to those categories. Processes
would then develop to serve those functions.

Finally, social practices can influence thinking habits
directly. Dialectics and logic can both be seen as cognitive
tools developed to deal with social conflict. We would not
expect that people whose social existence is based on harmony would develop a tradition of confrontation or
debate. On the contrary, when confronted with a conflict
of views, they might be oriented toward resolving the contradiction, transcending it, or finding a "Middle Way"—in
short, to approach matters dialectically. In contrast, people
who are free to argue might be expected to develop rules
for the conduct of debate, including the principle of noncontradiction and formal logic. It is an easy step from logic
to science, as physicist and historian of science Alan
Cromer has observed: "Science, in this view, is an extension of rhetoric. It was invented in Greece, and only in

Greece, because the Greek institution of the public assembly attached great prestige to debating skill. . . . A geometric proof is . . . the ultimate rhetorical form."

An important implication of this view of the causes of Greek and Chinese mental differences is the implied homeostasis. So long as economic forces operate to maintain different social structures, different social practices and child training will result in people focusing on different things in the environment. Focusing on different things will produce different understandings about the nature of the world. Different worldviews will in turn reinforce differential attention and social practices. The different worldviews will also prompt differences in perception and reasoning processes—which will tend to reinforce worldviews.

There is no reason to assume that the sequence ending in cognitive processes must begin with ecology. There can be many different economic reasons that might make some societies or groups more attentive to their fellow humans and many reasons that could make them more attentive to objects and their own goals with respect to them. For example, modern businesses and bureaucracies, and certainly entrepreneur-run businesses, do not necessarily require attention to a wide range of peers and a large number of supervisors. Instead, they require people to focus on a relatively narrow set of goals and to pursue them independently. Performance may actually be better if other people are largely ignored rather than attended to closely. The sequence need not even begin with economics. There can be many different reasons that could prompt attention to other people: for example, membership in a tightly knit religious community having strict rules for conduct. Similarly, many factors could cause people to focus primarily on objects and their goals with respect to them.

LATTER-DAY SUPPORT FOR THE ORIGIN THEORY

This economic-social account of cognition happens to fit with some important historical changes in the West. As the West became primarily agricultural in the Middle Ages, it became less individualistic. The European peasant was probably not much different from the Chinese peasant in terms of interdependence or freedom in daily life or in a rationalist approach to reasoning. And in terms of intellectual and cultural achievement, Europe had become a backwater. While Arab emirs discussed Plato and Aristotle and Chinese magistrates displayed their proficiency in all the arts, European nobles sat gnawing joints of beef in damp castles.

Toward the end of the Middle Ages, though, developments in European agriculture (notably the invention of the horse collar, which made possible the horse-drawn plow) created enough excess wealth that new trading centers, much like the old Greek city-states, appeared. The Italian city-states, and later the northern city-states, were to a very substantial degree autonomous and for the most part not subject to the authority of despots. Many of them also had a somewhat democratic, or at least oligarchic, character. And of course rebirth of the city-state form with its wealthy merchant class was associated with a renaissance of individualism, personal liberty, rationalism, and science. By the fifteenth century, Europe had awakened from its millennium of torpor and began to rival China in almost every domain—philosophy, mathematics, art, and technology.

An event that took place in the early fifteenth century is revealing about the differences between Europe and China. This was the voyage of the Grand Eunuch, on which hundreds of ships (technologically vastly superior to the *Pinta*, the *Niña*, and the *Santa Maria*) sailed from China to South and Southeast Asia, the Middle East, and Western Africa loaded with wealth and wonders. The voyage achieved its primary goal, which was to convince the nations bordering on the Indian Ocean, the Persian Gulf, and the Red Sea that China was superior in virtually every way to their own societies. But the Chinese were quite uninterested in seeing anything that those societies might have produced or known about—including even a giraffe that their African hosts showed them. The Chinese merely contended that the animal was known to them as a *qi lin*, a creature whose appearance was expected at the time of important events, such as the birth of a great emperor.

This lack of curiosity was characteristic of China. The inhabitants of the Middle Kingdom (China's name for itself, meaning essentially "the center of the world") had little interest in the tales brought to them by foreigners. Moreover, there has never been a strong interest in knowledge for its own sake in China. Even modern Chinese philosophers have always been far more interested in the pragmatic application of knowledge than with abstract theorizing for its own sake.

The intellectual advances that characterized Europe at
an increasing rate from the fifteenth century to the present seem to me to require more than an ecological or
geopolitical explanation of the sort offered by some recent
macrohistories, including Jared Diamond's brilliant *Guns,
Germs, and Steel*. While it is true that despotism and the
consequent suppression of opinion and initiative would
have been easier to carry off on ecological grounds in
China than in Europe, it seems to me to be a mistake to .
limit accounts of freedom of inquiry and scientific advance
in Europe to purely physical factors. Well before the fifteenth century, these values and the mentality that goes
with them had been implanted in the European mind.
Martin Luther launched his Ninety-Five Theses against
the abuses and tyranny of the Church not just because it
was easy for him to get away with it geographically, but
because the history of Europe had created a new sort of
person—one who conceived of individuals as separate
from the larger community and who thought in terms
imbued with freedom. Galileo and Newton made their
discoveries not just because they could not be readily sup-
pressed, but because of their curiosity and critical habits of
mind.

Now of course the East is drawing on the Western
stockpile of ideas at an ever-increasing rate. What effect
these ideas will have on the East, what they will look like
after being passed through an Eastern filter, and which
modifications may be adopted by the West can be guessed
at by looking at differences in the mental habits of contemporaries.

∝

As history, the account I am proposing for why Greece and China diverged as they did is speculative. It is nevertheless a scientific theory—because it leads to predictions that can be tested, and tested moreover in the psychological laboratory.

Twentieth-century psychologists have provided evidence that economic and social factors can affect perceptual habits. Herman Witkin and his colleagues showed that some people are less likely than others to separate an object from its surrounding environment. They called their dimension "field dependence"—referring to the degree to which perception of an object is influenced by the background or environment in which it appears. Witkin and his colleagues measured field dependence in a variety of ways. One of these was the Rod and Frame Test. In this test the participant looks into a long box at the end of which is a rod with a frame around it. The rod and frame can be tilted independently of each other and the participant's task is to indicate when the rod is completely vertical. The participant is accounted field dependent to the extent that judgments of the rod's verticality are influenced by the position of the frame. A second way of testing field dependence is to place people in a chair that tilts independently of the room in which it's placed. In this test, called the Body Adjustment Test, the participant is accounted field dependent to the extent that judgments of the verticality of the participant's own body are influenced by the tilt of the room. A third way, and the easiest to work with, is the Embedded Figures Test. In this test, the job is to locate a simple figure that is embedded in a much more complex figure. The longer it takes people to find the simple figure in its complicated context, the more field dependent they are assumed to be.

An implication of the idea that economic factors can affect cognitive habits is that agricultural peoples should be more field dependent than people who earn their living in ways that rely less on close coordination of their work with others, such as hunting animals and gathering plants. And in fact this is the case. We might also expect that traditional farming peoples would be more field dependent than people living in industrial societies, where personal goals can be pursued without close attention to a network of social roles and obligations. And this is also the case. In fact, hunter-gatherers and people in industrial societies are about equally field dependent.

If the key difference between agricultural peoples on the one hand and hunter-gatherers and modern, independent citizens of modern industrial societies on the other has to do with degree of attention to the social world, then it would be reasonable to expect that subcultures within a given society that differ in degree of social constraint should differ in degree of field dependence, as well. To test this hypothesis, personality psychologist Zachary Dershowitz examined the field dependence of Orthodox Jewish boys, who, he argued, live in families and social settings where role relations are spelled out quite explicitly and social constraints are substantial. He compared their performance with that of secular Jewish boys, who, he maintained, are subject to more lax social controls, and to that of Protestant boys, who, he believed, were exposed to even looser constraints. As expected, Dershowitz found the Orthodox boys to be more field dependent than the

secular Jewish boys, who in turn were more field dependent than the Protestant boys.

There is no reason to assume that field dependence
can only be the result of social constraints imposed from
the outside. We might expect that interest in other people,
whatever its origin, would be associated with field dependence. And in fact relatively field dependent people like to
be with other people more than relatively field independent people do. Field dependent people also have better
memory for faces and for social words ("visit," "party")
than relatively field independent people do. And, when
given their choice, field dependent people like to sit closer
to others than relatively field independent people do.

IMPLICATIONS FOR THOUGHT IN THE MODERN WORLD

But the implications of the view I am proposing extend far
beyond the confines of a particular style of perceiving
objects in relation to the environment. If something like
my account of the relation between social factors and
thought processes is correct—and if the social differences
between East and West today resemble those of ancient
times—then we can make some rather dramatic predictions about cognitive differences between contemporary
East Asians and Westerners. We might expect to find differences in:

- Patterns of attention and perception, with Easterners attending more to environments and

Westerners attending more to objects, and Easterners being more likely to detect relationships
among events than Westerners.

- Basic assumptions about the composition of the
  world, with Easterners seeing substances where
  Westerners see objects.
- Beliefs about controllability of the environment,
  with Westerners believing in controllability more
  than Easterners.
- Tacit assumptions about stability vs. change, with
  Westerners seeing stability where Easterners see
  change.
- Preferred patterns of explanation for events, with
  Westerners focusing on objects and Easterners
  casting a broader net to include the environment.
- Habits of organizing the world, with Westerners
  preferring categories and Easterners being more
  likely to emphasize relationships.
- Use of formal logical rules, with Westerners being
  more inclined to use logical rules to understand
  events than Easterners.
- Application of dialectical approaches, with Easterners being more inclined to seek the Middle
  Way when confronted with apparent contradiction and Westerners being more inclined to insist
  on the correctness of one belief vs. another.

At any rate, these are the expectations about habits of
mind that follow if it is really the case that Easterners and
Westerners have fundamentally different ways of seeing
themselves and the social world.

CHAPTER 3

LIVING TOGETHER VS. GOING IT ALONE

Most Westerners, or at any rate most Americans, are confident that the following generalizations apply to pretty much everyone:

- Each individual has a set of characteristic, distinctive attributes. Moreover, people *want* to be distinctive—different from other individuals in important ways.
- People are largely in control of their own behavior; they feel better when they are in situations in which choice and personal preference determine outcomes.
- People are oriented toward personal goals of success and achievement; they find that relation-

ships and group memberships sometimes get in the way of attaining these goals.

- People strive to feel good about themselves; personal successes and assurances that they have positive qualities are important to their sense of well-being.
- People prefer equality in personal relations or, when relationships are hierarchical, they prefer a superior position.
- People believe the same rules should apply to everyone—individuals should not be singled out for special treatment because of their personal attributes or connections to important people. Justice should be blind.

There are indeed hundreds of millions of such people, but they are to be found primarily in Europe, especially northern Europe, and in the present and former nations of the British Commonwealth, including the United States. The social-psychological characteristics of most of the rest of the world's people, especially those of East Asia, tend to be different to one degree or another.

THE NON-WESTERN SELF

There is an Asian expression that reflects a cultural prejudice against individuality: "The peg that stands out is pounded down." In general, East Asians are supposed to be less concerned with personal goals or self-aggrandizement than are Westerners. Group goals and coordinated action are more often the concerns. Maintaining harmonious social relations is likely to take precedence over achieving personal success. Success is often sought as a group goal rather than as a personal badge of merit. Individual distinctiveness is not particularly desirable. For Asians, feeling good about themselves is likely to be tied to the sense that they are in harmony with the wishes of the groups to which they belong and are meeting the group's expectations. Equality of treatment is not assumed nor is it necessarily regarded as desirable. 和平共处五项原则

The rules that apply to relationships in East Asia are presumed to be local, particular, and well specified by roles rather than universal. An Asian friend told me the most remarkable thing about visiting American households is that everyone is always thanking everyone else: "Thank you for setting the table"; "Thank you for getting the car washed." In her country everyone has clear obligations in a given context and you don't thank people for carrying out their obligations. Choice is not a high priority for most of the world's people. (An East Asian friend once asked me why Americans found it necessary to have a choice among forty breakfast cereals in the supermarket.) And Asians do not necessarily feel their competence as a decision maker is on the line when they do have to make a choice.

Most Americans over a certain age well remember their primer, called *Dick and Jane*. Dick and Jane and their dog, Spot, were quite the active individualists. The first page of an early edition from the 1930s (the primer was widely used until the 1960s) depicts a little boy running across a lawn. The first sentences are "See Dick run. See Dick play. See Dick run and play." This would seem the most natural sort of basic information to convey about kids—to the Western mentality. But the first page of the Chinese primer of the same era shows a little boy sitting on the shoulders of a bigger boy. "Big brother takes care of little brother. Big brother loves little brother. Little brother loves big brother." It is not individual action but relationships between people that seem important to convey in a child's first encounter with the printed word.

Indeed, the Western-style self is virtually a figment of the imagination to the East Asian. As philosopher Hu Shih writes, "In the Confucian human-centered philosophy man cannot exist alone; all action must be in the form of interaction between man and man." The person always exists within settings—in particular situations where there are particular people with whom one has relationships of a particular kind—and the notion that there can be attributes or actions that are not conditioned on social circumstances is foreign to the Asian mentality. Anthropologist Edward T. Hall introduced the notion of "low-context" vs. "high-context" societies to capture differences in self-understanding. To the Westerner, it makes sense to speak of a person as having attributes that are independent of circumstances or particular personal relations. This self— this bounded, impermeable free agent—can move from group to group and setting to setting without significant alteration. But for the Easterner (and for many other peoples to one degree or another), the person is connected, fluid, and conditional. As philosopher Donald Munro put it, East Asians understand themselves "in terms of their relation to the whole, such as the family, society, Tao Principle, or Pure Consciousness." The person participates in a set of relationships that make it possible to act and purely independent behavior is usually not possible or really even desirable.

Since all action is in concert with others, or at the very least affects others, harmony in relationships becomes a chief goal of social life. I have presented a schematic illustration intended to capture the different types of sense of self in relation to in-group, or close circle of friends and family; the illustration also conveys relative distance between in-group and out-group, or people who are mere acquaintances at most. Easterners feel embedded in their in-groups and distant from their out-groups. They tend to feel they are very similar to in-group members and they are much more trusting of them than of out-group members. Westerners feel relatively detached from their in-groups and tend not to make as great distinctions between in-group and out-group.

Some linguistic facts illustrate the social-psychological gap between East and West. In Chinese there is no word for "individualism." The closest one can come is the word for "selfishness." The Chinese character *jên*—benevolence—means two men. In Japanese, the word "I"—meaning the trans-situational, unconditional, generalized self with all its attributes, goals, abilities, and preferences—is not often used in conversation. Instead, Japanese has many words for "I," depending on audience and context. When a Japanese woman gives an official speech, she customarily uses *Watashi*, which is the closest Japanese comes to the trans-situational "I." When a man refers to himself in rela-

Eastern View

Western View

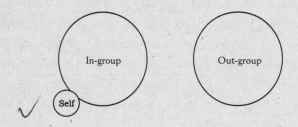

Eastern and Western views of the relations among self,
in-group, and out-group.

tion to his college chums he might say *Boku* or *Ore*. When
a father talks to his child, he says *Otosan* (Dad). A young
girl might refer to herself by her nickname when talking to
a family member: "Tomo is going to school today." The
Japanese often call themselves *Jibun*, the etymology of
which leads to a term meaning "my portion."

In Korean, the sentence "Could you come to dinner?"
requires different words for "you," which is common in

many languages, but also for "dinner," depending on whether one was inviting a student or a professor. Such practices reflect not mere politeness or self-effacement, but rather the Eastern conviction that one is a different person when interacting with different people.

"Tell me about yourself" seems a straightforward enough question to ask of someone, but the kind of answer you get very much depends on what society you ask it in. North Americans will tell you about their personality traits ("friendly, hard-working"), role categories ("teacher," "I work for a company that makes microchips"), and activities ("I go camping a lot"). Americans don't condition their self-descriptions much on context. The Chinese, Japanese, and Korean self, on the other hand, very much depends on context ("I am serious at work"; "I am fun-loving with my friends"). A study asking Japanese and Americans to describe themselves either in particular contexts or without specifying a particular kind of situation showed that Japanese found it very difficult to describe themselves without specifying a particular kind of situation—at work, at home, with friends, etc. Americans, in contrast, tended to be stumped when the investigator specified a context—"I am what I am." When describing themselves, Asians make reference to social roles ("I am Joan's friend") to a much greater extent than Americans do. Another study found that twice as many Japanese as American self-descriptions referred to other people ("I cook dinner with my sister").

When North Americans are surveyed about their attributes and preferences, they characteristically overestimate their distinctiveness. On question after question, North Americans report themselves to be more unique than they really are, whereas Asians are much less likely to make this error. Westerners also prefer uniqueness in the environment and in their possessions. Social psychologists Hee-jung Kim and Hazel Markus asked Koreans and Americans to choose which object in a pictured array of objects they preferred. Americans chose the rarest object, whereas Koreans chose the most common object. Asked to choose a pen as a gift, Americans chose the least common color offered and East Asians the most common.

It's revealing that the word for self-esteem in Japanese is *serufu esutiimu*. There is no indigenous term that captures the concept of feeling good about oneself. Westerners are more concerned with enhancing themselves in their own and others' eyes than are Easterners. Americans are much more likely to make spontaneous favorable comments about themselves than are Japanese. When self-appraisal measures are administered to Americans and Canadians, it turns out that, like the children of Lake Wobegon, they are pretty much all above average. Asians rate themselves much lower on most dimensions, not only endorsing fewer positive statements but being more likely to insist that they have negative qualities. It's not likely that the Asian ratings merely reflect a requirement for greater modesty than exists for North Americans. Asians are in fact under greater compunction to appear modest, but the difference in self-ratings exists even when participants think their answers are completely anonymous.

It isn't that Asians feel badly about their own attributes. Rather, there is no strong cultural obligation to feel that they are special or unusually talented. The goal for the self in relation to society is not so much to establish superiority or uniqueness, but to achieve harmony within a network of supportive social relationships and to play one's part in achieving collective ends. These goals require a certain amount of self-criticism—the opposite of tooting one's own horn. If I am to fit in with the group, I must root out those aspects of myself that annoy others or make their tasks more difficult. In contrast to the Asian practice of teaching children to blend harmoniously with others, some American children go to schools in which each child gets to be a "VIP" for a day. (In my hometown a few years ago the school board actually debated whether the chief goal of the schools should be to impart knowledge or to inculcate self-esteem. I appreciated a cartoon that appeared at about the same time showing a door with the label "Esteem Room.")

Japanese schoolchildren are taught how to practice self-criticism both in order to improve their relations with others and to become more skilled in solving problems. This stance of perfectionism through self-criticism continues throughout life. Sushi chefs and math teachers are not regarded as coming into their own until they've been at their jobs for a decade. Throughout their careers, in fact, Japanese teachers are observed and helped by their peers to become better at their jobs. Contrast this with the American practice of putting teachers' college graduates into the classroom after a few months of training and then leaving them alone to succeed or not, to the good or ill fortune of a generation of students.

An experiment by Steven Heine and his colleagues captures the difference between the Western push to feel good about the self and the Asian drive for self-improvement. The experimenters asked Canadian and Japanese students to take a bogus "creativity" test and then gave the students "feedback" indicating that they had done very well or very badly. The experimenters then secretly observed how long the participants worked on a similar task. The Canadians worked longer on the task if they had succeeded; the Japanese worked longer if they failed. The Japanese weren't being masochistic. They simply saw an opportunity for self-improvement and took it. The study has intriguing implications for skill development in both the East and West. Westerners are likely to get very good at a few things they start out doing well to begin with. Easterners seem more likely to become Jacks and Jills of all trades.

INDEPENDENCE VS. INTERDEPENDENCE

The broad differentiation between the two types of societies we have been discussing has been a staple notion of social science since the nineteenth century. The distinction is similar to that made by nineteenth-century German social scientists, notably Ferdinand Tönnies, who made a useful distinction for comparing cultures, namely between a *Gemeinschaft* (a community based on a shared sense of identity) and a *Gesellschaft* (an institution intended to facilitate action to achieve instrumental goals). A *Gemeinschaft* is based on relationships that exist for their own sake and rest on a sense of unity and mutuality: for example, relationships among family members, church congregations, or a network of friends. It is based on sympathy, frequent face-to-face interaction, shared experiences, and even shared property. A *Gesellschaft* is based on interactions that are mostly a means to an end. It frequently involves exchange of goods and labor and is often based on bargaining and contracts. Such social systems allow for personal gain and competitive advantage. Corporations and bureaucracies are examples of *Gesellschaften*.

No one thinks a given institution or society is exclusively of the *Gemeinschaft* or *Gesellschaft* sort. They are merely ideal types. But the distinction is of great analytic importance for much of modern social science, especially for cultural psychology. The *Gemeinschaft* is often termed a "collectivist" social system and the *Gesellschaft* is often labeled an "individualist" social system. The terms "interdependent" and "independent," proposed by Hazel Markus and Shinobu Kitayama, convey similar notions, and these are the ones I will normally use.

Training for independence or interdependence starts quite literally in the crib. Whereas it is common for American babies to sleep in a bed separate from their parents, or even in a separate room, this is rare for East Asian babies—and, for that matter, babies pretty much everywhere else. Instead, sleeping in the same bed is far more common. The differences are intensified in waking life. Adoring adults from several generations often surround the Chinese baby (even before the one-child policy began producing "little emperors"). The Japanese baby is almost always with its mother. The close association with mother is a condition that some Japanese apparently would like to continue indefinitely. Investigators at the University of Michigan's Institute for Social Research recently conducted a study requiring a scale comparing the degree to which adult Japanese and American respondents want to be with their mothers. The task proved very difficult, because the Japanese investigators insisted that a reasonable endpoint on the scale would be "I want to be with my mother almost all the time." The Americans, of course, insisted that this would be uproariously funny to American respondents and would cause them to cease taking the interview seriously.

Independence for Western children is often encouraged in rather explicit ways. Western parents constantly require their children to do things on their own and ask them to make their own choices. "Would you like to go to bed now or would you like to have a snack first?" The Asian parent makes the decision for the child on the assumption that the parent knows best what is good for the child.

Parents who work to create an independent child shouldn't be surprised when the training works so well that their children balk at threats to their freedom of choice. Social psychologists Sheena Iyengar and Mark Lepper asked American, Chinese, and Japanese children aged seven to nine to solve anagrams, such as, "What word can be made from GREIT?" Some of the children were told to work on a particular category of anagrams; other children were given a choice about which anagrams to solve; and still others were told that the experimenter had spoken to the child's mom, who would like the child to work on a particular category. The researchers then measured the number of anagrams solved and the time spent working on them. The American children showed the highest level of motivation—spending more time on the task and solving more anagrams—when they were allowed to choose the category. The American children showed the least motivation when it was Mom who chose the category, suggesting that they felt their autonomy had been encroached upon and they had therefore lost some of their intrinsic interest in the task. The Asian children showed the highest level of motivation when Mom chose the category.

An emphasis on relationships encourages a concern with the feelings of others. When American mothers play with their toddlers, they tend to ask questions about objects and supply information about them. But when Japanese mothers play with their toddlers, their questions are more likely to concern feelings. Japanese mothers are particularly likely to use feeling-related words when their children misbehave: "The farmer feels bad if you did not eat everything your mom cooked for you." "The toy is crying because you threw it." "The wall says 'ouch.' " Concentrating attention on objects, as American parents tend to do, helps to prepare children for a world in which they are expected to act independently. Focusing on feelings and social relations, as Asian parents tend to do, helps children to anticipate the reactions of other people with whom they will have to coordinate their behavior.

The consequences of this differential focus on the emotional states of others can be seen in adulthood. There is evidence that Asians are more accurately aware of the feelings and attitudes of others than are Westerners. For example, Jeffrey Sanchez-Burks and his colleagues showed to Koreans and Americans evaluations that employers had made on rating scales. The Koreans were better able to infer from the ratings just what the employers felt about their employees than were the Americans, who tended to simply take the ratings at face value. This focus on others' emotions extends even to perceptions of the animal world. Taka Masuda and I showed underwater video scenes to Japanese and American students and asked them to report what they saw. The Japanese students reported "seeing" more feelings and motivations on the part of fish than did Americans; for example, "The red fish must be angry because its scales were hurt." Similarly, Kaiping Peng and Phoebe Ellsworth showed Chinese and American students animated pictures of fish moving in various patterns in relation to one another. For example, a group might appear to chase an individual fish or to scoot away when the individual fish approached. The investigators asked the students what both the individual fish and the groups of fish were feeling. The Chinese readily complied with the requests. The Americans had difficulty with both tasks and were literally baffled when asked to report what the group emotions might be.

The relative degree of sensitivity to others' emotions is reflected in tacit assumptions about the nature of communication. Westerners teach their children to communicate their ideas clearly and to adopt a "transmitter" orientation, that is, the speaker is responsible for uttering sentences that can be clearly understood by the hearer—and understood, in fact, more or less independently of the context. It's the speaker's fault if there is a miscommunication. Asians, in contrast, teach their children a "receiver" orientation, meaning that it is the hearer's responsibility to understand what is being said. If a child's loud singing annoys an American parent, the parent would be likely just to tell the kid to pipe down. No ambiguity there. The Asian parent would be more likely to say, "How well you sing a song." At first the child might feel pleased, but it would likely dawn on the child that something else might have been meant and the child would try being quieter or not singing at all.

Westerners—and perhaps especially Americans—are apt to find Asians hard to read because Asians are likely to assume that their point has been made indirectly and with finesse. Meanwhile, the Westerner is in fact very much in the dark. Asians, in turn, are apt to find Westerners—perhaps especially Americans—direct to the point of condescension or even rudeness.

There are many ways of parsing the distinction between relatively independent and relatively interdependent societies, but in illustrating these it may be helpful to focus on four related but somewhat distinct dimensions:

- Insistence on freedom of individual action vs. a preference for collective action.
- Desire for individual distinctiveness vs. a preference for blending harmoniously with the group.
- A preference for egalitarianism and achieved status vs. acceptance of hierarchy and ascribed status.
- A belief that the rules governing proper behavior should be universal vs. a preference for particularistic approaches that take into account

the context and the nature of the relationships involved.

These dimensions are merely correlated with one another; and it is possible, for example, for a given society to be quite independent in terms of some dimensions and much less so in terms of others. Social scientists have attempted to measure each of these dimensions, and other associated ones, in a variety of ways, including value surveys, studies of archived material, and experiments.

Some of the most interesting survey material comes from the study of businesspeople from different cultures. Such surveys provide particularly convincing evidence because so much is held more or less constant, including relative wealth and educational levels. In the classic study of this sort, Geert Hofstede provided even more comparability than that: All of his participants, who came from dozens of different societies, were employees of IBM. He found dramatic cultural differences in values even among Big Blue employees.

Similar data have been collected by Charles Hampden-Turner and Alfons Trompenaars, who are professors at an international business school in Holland. Over a period of several years they gave dozens of questions to middle managers taking seminars they conduct throughout the world. The participants in their seminars—fifteen thousand all told—were from the U.S., Canada, Australia, Britain, the Netherlands, Sweden, Belgium, Germany, France, Italy, Singapore, and Japan (and a small number from Spain and Korea, as well). Hampden-Turner and Trompenaars presented their students with dilemmas in which independent values were pitted against interdependent values.

To examine the value of individual distinction vs. harmonious relations with the group, Hampden-Turner and Trompenaars asked the managers to indicate which of the following types of job they preferred: (a) jobs in which personal initiatives are encouraged and individual initiatives are achieved; versus (b) jobs in which no one is singled out for personal honor, but in which everyone works together.

More than 90 percent of American, Canadian, Australian, British, Dutch, and Swedish respondents endorsed the first choice—the individual freedom alternative—vs. fewer than 50 percent of Japanese and Singaporeans. Preferences of the Germans, Italians, Belgians, and French were intermediate.

The U.S. is sometimes described as a place where, if you claim to amount to much, you should be able to show that you change your area code every five years or so. (This was before the phone company started changing people's area codes without waiting for them to move.) In some other countries, the relationship with the corporation where one is employed, and the connection with one's colleagues there, are more highly valued than in the U.S. and presumed to be more or less permanent. To assess this difference among cultures, Hampden-Turner and Trompenaars asked their participants to choose between the following expectations: If I apply for a job in a company, (a) I will almost certainly work there for the rest of my life; or (b) I am almost sure the relationship will have a limited duration.

More than 90 percent of Americans, Canadians, Australians, British, and Dutch thought a limited job duration was likely. This was true for only about 40 percent of Japanese (though it would doubtless be substantially higher today after "downsizing" has come even to Japan). The French, Germans, Italians, and Belgians were again intermediate, though closer to the other Europeans than to the Asians.

To examine the relative value placed on achieved vs. ascribed status, Hampden-Turner and Trompenaars asked their participants whether or not they shared the following view: Becoming successful and respected is a matter of hard work. It is important for a manager to be older than his subordinates. Older people should be more respected than younger people.

More than 60 percent of American, Canadian, Australian, Swedish, and British respondents rejected the idea of status being based in any way on age. About 60 percent of Japanese, Korean, and Singapore respondents accepted hierarchy based in part on age; French, Italians, Germans, and Belgians were again intermediate, though closer to the other Europeans than to the Asians.

Needless to say, there is great potential for conflict when people from cultures having different orientations must deal with one another. This is particularly true when people who value universal rules deal with people who think each particular situation should be examined on its merits and that different rules might be appropriate for different people. Westerners prefer to live by abstract principles and like to believe these principles are applicable to everyone. To set aside universal rules in order to accommodate particular cases seems immoral to the Westerner. To insist on the same rules for every case can seem at best obtuse and rigid to the Easterner and at worst cruel. Many of Hampden-Turner and Trompenaars's questions reveal what a marked difference exists among cultures in their preference for universally applicable rules vs. special consideration of cases based on their distinctive aspects. One of their questions deals with how to handle the case of an employee whose work for a company, though excellent for fifteen years, has been unsatisfactory for a year. If there is no reason to expect that performance will improve, should the employee be (a) dismissed on the grounds that job performance should remain the grounds for dismissal, regardless of the age of the person and his previous record; or (b) is it wrong to disregard the fifteen years the employee has been working for the company? One has to take into account the company's responsibility for his life.

More than 75 percent of Americans and Canadians felt the employee should be let go. About 20 percent of Koreans and Singaporeans agreed with that view. About 30 percent of Japanese, French, Italians, and Germans agreed and about 40 percent of British, Australians, Dutch, and Belgians agreed. (Atypically for this question, the British and the Australians were closer to the continental Europeans than to the North Americans.)

As these results show, Westerners' commitment to universally applied rules influences their understanding of the nature of agreements between individuals and between corporations. By extension, in the Western view, once a contract has been agreed to, it is binding—regardless of circumstances that might make the arrangement much less attractive to one of the parties than it had been initially. But to people from interdependent, high-context cultures, changing circumstances dictate alterations of the agreement.

These very different outlooks regularly produce international misunderstandings. The Japanese-Australian "sugar contract" case in the mid-1970s provides a particularly dramatic example. Japanese sugar refiners contracted with Australian suppliers to provide them with sugar over a period of five years at the price of $160 per ton. But shortly after the contract was signed, the value of sugar on the world market dropped dramatically. The Japanese thereupon asked for a renegotiation of the contract on the grounds that circumstances had changed radically. But to the Australians, the agreement was binding, regardless of circumstances, and they refused to consider any changes.

An important business implication of the differences that exist between independent and interdependent societies is that advertising needs to be modified for particular cultural audiences. Marketing experts Sang-pil Han and Sharon Shavitt analyzed American and Korean advertisements in popular news magazines and women's magazines. They found that American advertisements emphasize individual benefits and preferences ("Make your way through the crowd"; "Alive with pleasure"), whereas Korean advertisements are more likely to emphasize collective ones ("We have a way of bringing people closer together"; "Ringing out the news of business friendships that really work"). When Han and Shavitt performed experiments, showing people different kinds of advertisements, they found that the individualist advertisements were more effective with Americans and the collectivist ones with Koreans.

Independence vs. interdependence is of course not an either/or matter. Every society—and every individual—is a blend of both. It turns out that it is remarkably easy to bring one or another orientation to the fore. Psychologists Wendi Gardner, Shira Gabriel, and Angela Lee "primed" American college students to think either independently or interdependently. They did this in two different ways. In one experiment, participants were asked to read a story about a general who had to choose a warrior to send to the king. In an "independent" version, the king had to choose the best individual for the job. In an "interdependent" version the general wanted to make a choice that would benefit his family. In another priming method, participants were asked to search for words in a paragraph describing a trip to a city. The words were either independent in nature (e.g., "I," "mine") or interdependent (e.g., "we," "ours").

After reading the story or searching for words in the paragraph, participants were asked to fill out a value survey that assessed the importance they placed on individualist values (such as freedom and living a varied life) and collectivist values (such as belongingness and respect for elders). They also read a story in which "Lisa" refused to give her friend "Amy" directions to an art store because she was engrossed in reading a book; they were then asked whether Lisa's behavior was inappropriately selfish. Students who had been exposed to an independence prime rated individualist values higher and collectivist values lower than did students exposed to an interdependence prime. The independence-primed participants were also more forgiving of the book-engrossed Lisa. Gardner and her colleagues repeated their study adding Hong Kong students to their American sample and also added an unprimed control condition. American students rated individualist values higher than collectivist values—unless they had been exposed to an interdependence prime. Hong Kong students rated collectivist values higher than individualist values—unless they had been exposed to an independence prime.

Of course, Easterners are constantly being "primed" with interdependence cues and Westerners with independence cues. This raises the possibility that even if their upbringing had not made them inclined in one direction or another, the cues that surround them would make people living in interdependent societies behave in generally interdependent ways and those living in independent societies behave in generally independent ways. In fact this is a common report of people who live in the "other" culture for a while. My favorite example concerns a young Canadian psychologist who lived for several years in Japan. He then applied for jobs at North American universities. His adviser was horrified to discover that his letter began with apologies about his unworthiness for the jobs in question. Other evidence shows that self-esteem is highly malleable. Japanese who live in the West for a while show a notable increase in self-esteem, probably because the situations they encountered were in general more esteem-enhancing than those typical in Japan. The social psychological characteristics of people raised in very different cultures are far from completely immutable.

VARIANTS OF VIEWPOINT

The work of Hampden-Turner and Trompenaars makes clear that the West is no monolith concerning issues of independence vs. interdependence. There are also substantial regularities to the differences found in Western countries. In general, the Mediterranean countries plus Belgium and Germany are intermediate between the East Asian countries on the one hand and the countries most heavily influenced by Protestant, Anglo-Saxon culture on the other. There is more regularity even than that. Someone has said, "The Idea moves west," meaning that the values of individuality, freedom, rationality, and universalism became progressively more dominant and articulated as civilization moved westward from its origins in the Fertile Crescent. The Babylonians codified and universalized the law. The Israelites emphasized individual distinctiveness. The Greeks valued individuality even more and added a commitment to personal freedom, the spirit of debate, and formal logic. The Romans brought a gift for rational organization and something resembling the Chinese genius for technological achievement, and—after a trough lasting almost a millennium—their successors, the Italians, rediscovered these values and built on the accomplishments of the Greek and Roman eras. The Protestant Reformation, beginning in Germany and Switzerland and largely bypassing France and Belgium, added individual responsibility and a definition of work as a sacred activity. The Reformation also brought a weakened commitment to the family and other in-groups coupled with a greater willingness to trust out-groups and have dealings with their members.

These values were all intensified in the Calvinist subcultures of Britain, including the Puritans and Presbyterians, whose egalitarian ideology laid the groundwork for the government of the United States. (Thomas Jefferson was merely paraphrasing the Puritan sympathizer John Locke when he wrote, "We hold these truths to be self-evident, that all men are created equal . . . with certain inalienable rights, that among these are life, liberty. . . .")

The Hampden-Turner and Trompenaars findings for social values, as well as those of Hofstede, track this East-West ideological journey almost exactly. The further to the West a given country lies, the greater, in general, that country's endorsement of independent values. Moreover, these differences among European cultures are reflected in their successor subcultures in the United States, a fact documented in immigrant cultural histories by scholars such as economist Thomas Sowell. I once knew a very distinguished and well-placed social scientist, a crusty Scottish-American Presbyterian steeped in Calvinist rectitude. He had a son who was also a social scientist and who had to struggle from time to time to sustain his career during the 1970s, when jobs were scarce in the U.S. My colleague would sometimes state proudly that, although it would have been easy for him to do so, he had never intervened to help his son's situation. The colleague's Anglo-Saxon Protestant friends would nod their approval of the justice of this stance in the face of the personal pain they knew the colleague had experienced. His Jewish and Catholic colleagues, with their more Continental values, would stare in shocked disbelief at his lack of family feeling. At a level slightly more scientific than this anecdote: We generally find that it is the white Protestants among the American participants in our studies who show the most "Western" patterns of behavior and that Catholics and minority group members, including African Americans and Hispanics, are shifted somewhat toward Eastern patterns.

There are also major differences among Eastern cultures in all sorts of important social behavior and values, some of which are related to independence versus interdependence.

I was in China in 1982 at the tail end of the Cultural Revolution. The country seemed extremely exotic—in both its traditional aspects and its Communist-imposed aspects. (This was well before a Starbucks was installed in the Forbidden City!) The first Western play to be performed in Beijing since the revolution was mounted while I was there. It was Arthur Miller's *Death of a Salesman*. The choice seemed very strange. I regarded the play as being not merely highly Western in character but distinctly American. Its central figure is a salesman, "a man way out there in the blue riding on a smile and a shoeshine." To my astonishment, the play was a tremendous success. But Arthur Miller, who had come to China to collaborate on production of the play, provided a satisfactory reason for its reception. "The play is about family," he said, "and the Chinese invented family." He might have added that the play is also about *face*, or the need to have the respect of the community, and the Chinese also invented face.

The Japanese have perhaps as much concern with face as do the Chinese, but probably less involvement with the

immediate family and more commitment to the corporation. There are other marked differences between the
Japanese and Chinese. The sociologist Robert Bellah, the
philosopher Hajime Nakamura, the psychologist Dora
Dien, and the social philosopher Lin Yutang, among many
others, have detailed some of these differences. Though
social constraints are in general greater on both Chinese
and Japanese than on Westerners, the constraints come
primarily from authorities in the case of the Chinese and
chiefly from peers in the case of the Japanese. Control in
Chinese classrooms, for example, is achieved by the
teacher, but by classmates in Japan. Dora Dien has written
that the "Chinese emphasize particular dyadic [two-
person] relationships while retaining their individuality,
whereas the Japanese tend to submerge themselves in the
group." Though both Chinese and Japanese are required to
conform to move smoothly through their daily lives, the
Chinese are said to chafe under the requirements and the
Japanese actually to enjoy them. The Japanese are held to
share with the Germans and the Dutch a need for order in
all spheres of their lives; the Chinese share with Mediter-
raneans a more relaxed approach to life.

It is sometimes argued that one particular type of
social relationship is unique to the Japanese. This is *amae*,
a concept discussed at length by the Japanese psychoana-
lyst Takeo Doi. *Amae* describes a relationship in which an
inferior, a child or employee, for example, is allowed to
engage in inappropriate behavior—to ask for an expensive
toy or to request a promotion at a time not justified by
company policy—as an expression of confidence that the
relationship is sufficiently close that the superior will be

indulgent. *Amae* facilitates the relationship, enhancing trust between the two parties and cementing bonds, though these results come at some cost to the autonomy of the inferior.

The very real differences among Eastern cultures and among Western cultures, however, shouldn't blind us to the fact that the East and West are in general quite different from each other with respect to a great many centrally important values and social-psychological attributes.

AWASE AND *ERABI*

STYLES OF CONFLICT AND NEGOTIATION

Debate is almost as uncommon in modern Asia as in ancient China. In fact, the whole rhetoric of argumentation that is second nature to Westerners is largely absent in Asia. North Americans begin to express opinions and justify them as early as the show-and-tell sessions of nursery school ("This is my robot; he's fun to play with because . . ."). In contrast, there is not much argumentation or trafficking in opinions in Asian life. A Japanese friend has told me that the concept of a "lively discussion" does not exist in Japan—because of the risk to group harmony. It is this fact that likely undermined an attempt he once made to have an American-style dinner party in Japan, inviting only Japanese guests who expressed a fondness for the institution—from the martinis through the steak to the apple pie. The effort fell flat for want of opinions and people willing to defend them.

The absence of a tradition of debate has particularly dramatic implications for the conduct of political life. Very recently, South Korea installed its first democratic government. Prior to that, it had been illegal to discuss North Korea. Westerners find this hard to comprehend, inasmuch as South Korea has performed one of the world's most impressive economic miracles of the past 40 years and North Korea is a failed state in every respect. But, due to the absence of a tradition of debate, Koreans have no faith that correct ideas will win in the marketplace of ideas, and previous governments "protected" their citizens by preventing discussion of Communist ideas and North Korean practices.

The tradition of debate goes hand in hand with a certain style of rhetoric in the law and in science. The rhetoric of scientific papers consists of an overview of the ideas to be considered, a description of the relevant basic theories, a specific hypothesis, a statement of the methods and justification of them, a presentation of the evidence produced by the methods, an argument as to why the evidence supports the hypothesis, a refutation of possible counterarguments, a reference back to the basic theory, and a comment on the larger territory of which the article is a part. For Americans, this rhetoric is constructed bit by bit from nursery school through college. By the time they are graduate students, it is second nature. But for the most part, the rhetoric is new to the Asian student and learning it can be a slow and painful process. It is not uncommon for American science professors to be impressed by their hard-working, highly selected Asian students and then to be disappointed by their first major paper—not because of their incomplete command of English, but because of their lack of mastery of the rhetoric common in the professor's field. In my experience, it is also not uncommon for professors to fail to recognize that it is the lack of the Western rhetoric style they are objecting to, rather than some deeper lack of comprehension of the enterprise they're engaged in.

The combative, rhetorical form is also absent from Asian law. In Asia the law does not consist, as it does in the West for the most part, of a contest between opponents. More typically, the disputants take their case to a middleman whose goal is not fairness but animosity reduction—by seeking a Middle Way through the claims of the opponents. There is no attempt to derive a resolution to a legal conflict from a universal principle. On the contrary, Asians are likely to consider justice in the abstract, by-the-book Western sense to be rigid and unfeeling.

Negotiation also has a different character in the high-context societies of the East than in the low-context societies of the West. Political scientist Mushakoji Kinhide characterizes the Western *erabi* (active, agentic) style as being grounded in the belief that "man can freely manipulate his environment for his own purposes. This view implies a behavioral sequence whereby a person sets his objective, develops a plan designed to reach that objective, and then acts to change the environment in accordance with that plan." To a person having such a style, there's not much point in concentrating on relationships. It's the results that count. Proposals and decisions tend to be of the either/or variety because the Westerner knows what he wants and has a clear idea what it is appropriate to give and to take in order to have an acceptable deal. Negotiations should be short and to the point, so as not to waste time reaching the goal.

The Japanese *awase* (harmonious, fitting-in) style, "rejects the idea that man can manipulate the environment and assumes instead that he adjusts himself to it." Negotiations are not thought of as "ballistic," one-shot efforts never to be revisited, and relationships are presumed to be long-term. Either/or choices are avoided. There is a belief that "short-term wisdom may be long-term folly." A Japanese negotiator may yield more in negotiations for a first deal than a similarly placed Westerner might, expecting that this will lay the groundwork for future trust and cooperation. Issues are presumed to be complex, subjective, and intertwined, unlike the simplicity, objectivity, and "fragmentability" that the American with the *erabi* style assumes.

So there are very dramatic social-psychological differences between East Asians as a group and people of European culture as a group. East Asians live in an interdependent world in which the self is part of a larger whole; Westerners live in a world in which the self is a unitary free agent. Easterners value success and achievement in good part because they reflect well on the groups they belong to; Westerners value these things because they are badges of personal merit. Easterners value fitting in and engage in self-criticism to make sure that they do so; Westerners value individuality and strive to make themselves look good. Easterners are highly attuned to the feelings of others and strive for interpersonal harmony; Westerners are more concerned with knowing themselves and are prepared to sacrifice harmony for fairness. Easterners are accepting of hierarchy and group control; Westerners are more likely to prefer equality and scope for personal action. Asians avoid controversy and debate; Westerners have faith in the rhetoric of argumentation in arenas from the law to politics to science.

None of these generalizations apply to all members of their respective groups, of course. Every society has individuals who more nearly resemble those of other, quite different societies than they do those of their own society; and every individual within a given society moves quite a bit between the independent and interdependent poles over the course of a lifetime—over the course of a day, in fact. But the variations between and within societies, as well as within individuals, should not blind us to the fact that there are very real differences, substantial on the average, between East Asians and people of European culture.

As nearly as we can tell, these social differences are much the same as the differences that characterized the ancient Chinese and Greeks. And if it was the social circumstances that produced the cognitive differences between ancient Chinese and Greeks, then we might expect to find cognitive differences between modern East Asians and Westerners that map onto the differences between the ancient Chinese and Greeks.

CHAPTER 4

"EYES IN BACK OF YOUR HEAD" OR "KEEP YOUR EYE ON THE BALL"?

If people really do see the world in terms dictated by their social existence, then we might expect modern East Asians to have the same sort of holistic worldviews as ancient Chinese thinkers did, and we might expect modern people of European culture to exhibit the same sorts of analytic approaches that were characteristic of ancient Greek thinkers. Moreover, the different social realities might produce very different patterns of literally *seeing* the world. People who live in a world in which external forces are the important ones could be expected to pay close attention to the environment. People who live in a world in which personal agency produces results might focus primarily on objects that they can manipulate to serve their own goals.

HOLISM VS. ANALYSIS

I was sitting on a plane bound from northern California recently when I heard the voice of a man—a European American—asking questions of his two-and-a-half-year-old son.

> Dad: "What shape is the balloon?" No answer. "It's round, Jason."
> Dad: "This is a pair of socks. Are they short or long?"
> Little boy: "Short."
> Dad: "That's right, short."
> Dad: "This is a pair of pants. Are they . . . ?"
> Little boy: "Short."
> Dad: "No, Jason, they're long."

Though this exchange may seem to Westerners to be an unexceptional quiz, by Asian standards it is quite unusual. The father's questions consisted of directing his son's attention to objects and asking about their properties. Whereas this might seem to Westerners to be the most natural way to orient a child's attention, it's not to Easterners, and the reasons for this have profound implications for cultural differences in perception and cognition.

The ancient Chinese philosophers saw the world as consisting of continuous substances and the ancient Greek philosophers tended to see the world as being composed of discrete objects or separate atoms. A piece of wood to the Chinese would have been a seamless, uniform material; to the Greeks it would have been seen as composed of particles. A novel item, such as a seashell, might have been seen as a substance by the Chinese and as an object by the Greeks. Remarkably, there is evidence that modern Asians also tend to see the world as consisting of continuous substances, whereas modern Westerners are more prone to see objects.

Cognitive psychologists Mutsumi Imae and Dedre Gentner showed objects composed of particular substances to Japanese and Americans of various ages from less than two to adulthood and described them in ways that were neutral with respect to whether each was an object or a substance. For example, they might show a pyramid made of cork and ask the participants to "look at this 'dax.' " Then they showed the participants two trays, one of which had something on it of the same shape as the object presented but which was made of a different substance (for example, a pyramid made of white plastic) and one of which had the same substance in a different shape (for example, pieces of cork). The investigators then asked their participants to point to the tray that had *their* "dax" on it.

Americans were much more likely to choose the same shape as the "dax" than were the Japanese, indicating that the Americans were coding what they saw as an object. The Japanese were more likely to choose the same material as the "dax," indicating that they were coding what they saw as a substance. The differences between Americans and Japanese were very large. On average, across the many trials with different displays, more than two thirds of four-year-old American children chose another object as the "dax," whereas fewer than a third of Japanese four-year-old children did. The differences were equally large for adults. Even two-year-olds differed. American infants were somewhat more likely to choose the object than were the Japanese infants.

Taken at face value, the Imai and Gentner results indicate that Westerners and Asians literally see different worlds. Like ancient Greek philosophers, modern Western-ers see a world of objects—discrete and unconnected *things*. Like ancient Chinese philosophers, modern Asians are inclined to see a world of substances—continuous masses of *matter*. The Westerner sees an abstract statue where the Asian sees a piece of marble; the Westerner sees a wall where the Asian sees concrete. There is much other evidence—of a historical, anecdotal, and systematic scien-tific nature—indicating that Westerners have an analytic view focusing on salient objects and their attributes, whereas Easterners have a holistic view focusing on conti-nuities in substances and relationships in the environment.

In the turn-of-the-century midwestern neighborhood where I live in Ann Arbor, Michigan, many of the homes are attractive workers' cottages with white clapboard sid-ing and gabled roofs. The homes were shipped by the Sears Roebuck Company and unloaded at the railroad sta-tion before being brought up the hill by horse carts to be put together like a puzzle from numbered pieces. Not too long after, Henry Ford, whose motor car company was and is located about forty miles from my town, invented the assembly line. Auto part "atoms" were put together by workers performing a repetitive, identical set of actions over and over again at a fixed station in the line. Iron ore came in one end of the River Rouge plant in Dearborn, Michigan, and, after being smelted and manufactured into simple parts and put together by workers performing simple operations, came out as a Model A Ford on the other.

Beginning in the late eighteenth and early nineteenth century, the West, and especially America, began to atomize, that is to say, *modularize* the worlds of manufacture and commerce. The production of everything from muskets to furniture was broken down into the most standardized parts possible and the simplest replicable actions. Each implement, each component, each action of the worker was analyzed and made maximally efficient. Objects that had taken craftsmen months to create could now be produced in a matter of hours. Time itself became a modular entity: three minutes for bolting on the carburetor, two and a half for installing spark plugs.

Starting around the late nineteenth century, retail stores became modular "chains." It was possible to go into a Sears and, a half century or so later, a McDonald's, anywhere in the country—and eventually the world—and see the same rows of merchandise, or the same booths and burgers, in any of them. (One of my favorite *New Yorker* cartoons depicts two older American ladies asking a hotel doorman, "Is this the Geneva Sheraton or the Brussels Sheraton?")

The atomistic attitude of Westerners extends to their understanding of the nature of social institutions. In their survey of the values of middle managers, Hampden-Turner and Trompenaars asked whether their respondents thought of a company as a system to organize tasks or as an organism coordinating people working together:

(a) A company is a system designed to perform
     functions and tasks in an efficient way. People

are hired to fulfill these functions with the help
of machines and other equipment. They are
paid for the tasks they perform.

(b) A company is a group of people working
together. The people have social relations with
other people and with the organization. The
functioning is dependent on these relations.

About 75 percent of Americans chose the first definition, more than 50 percent of Canadians, Australians,
British, Dutch, and Swedes chose that definition, and about
a third of Japanese and Singaporese chose it. Germans,
French, and Italians as a group were intermediate between
the Asians and the people of British and northern European culture. Thus for the Westerners, especially the Americans and the other people of primarily northern European
culture, a company is an atomistic, modular place where
people perform their distinctive functions. For the Easterners, and to a lesser extent the eastern and southern Europeans, a company is an organism where the social relations
are an integral part of what holds things together.

The holism of the ancient Chinese extended to a sense
of the unity of human existence with natural and even
supernatural occurrences. What happened on earth res-
onated with events in nature and in heaven. The same is
true of East Asians today. Both Taoism, still influential in
China and elsewhere in East Asia, and Shintoism, still
important in Japan, retain strong elements of animism:
animals, plants, natural objects, and even human-made
artifacts have spirits. Advertisements that emphasize

nature have far more success in Asia than in the West. The Nissan corporation discovered this fact, to its chagrin, when it opened its advertising campaign for the Infiniti luxury car in the U.S. not with pictures of its automobile but with scenes of nature—often several expensive pages of nature scenes in a row—with just the name of the car at the end of the sequence. The campaign was a noted flop. ("Although," quipped one American advertising industry wag, "sales of rocks and trees are way up.")

Just as the social attitudes and values of continental Europe are intermediate between East Asian and Anglo-American ones, the intellectual history of the Continent is more holistic than that of America and the Common-wealth. The big-picture ideas are much rarer in Anglo-America than on the Continent. During the many decades that Anglo-American philosophers concerned themselves with atomistic, so-called ordinary language analysis, Euro-pean philosophers were inventing phenomenology, existen-tialism, structuralism, poststructuralism, and postmodernism. The largest systems of political, economic, and social thought arise primarily from the Continent. Marxism is a German product; sociology was invented by the French-man Auguste Comte and raised to its highest level of achievement by the German Max Weber. In psychology, it is also the continentals who dominate the big-picture theo-ries: the Austrian Freud and the Swiss Piaget are perhaps the most influential psychologists of the twentieth century. In my own subfield of social psychology, two Germans, Kurt Lewin and Fritz Heider, have contributed by far the broadest and most comprehensive theories. And the school of psychology that I find myself belatedly belonging to is the historical-cultural one established by the Russian psychologists Lev Vygotsky and Alexander Luria.

It's not just that Anglo-American scholars don't tend to create broad-ranging theories; they can seem positively allergic to them. B. F. Skinner, America's chief candidate for the psychology pantheon, was not merely a reductionist of the extreme atomic school, he actually believed theories of any sort were inappropriate—too general and too removed from the unshakable facts. Students in my graduate school cohort who toyed with large ideas were likely to be accused by their peers of engaging in "night-school metaphysics." Even Anglo-American social scientists who are sympathetic to theories don't tend to like the big ones. My sociology teacher in graduate school was Robert Merton, who praised "theories of the middle range" as being the right level to aim for. (To his dismay, this was once translated by an Italian scholar, perhaps tongue in cheek, as "theories of the average level.")

PERCEIVING THE WORLD

If East Asians must coordinate their behavior with others and adjust to situations, we would expect them to attend more closely to other people's attitudes and behaviors than do Westerners. In fact we have evidence that East Asians do pay more attention to the social world than do Westerners. Li-jun Ji, Norbert Schwarz, and I found evidence that Beijing University students have more knowledge about the attitudes and behaviors of their peers than do University of Michigan students. A research team from our labs at Michigan headed by Trey Hedden and Denise Park, and by Qicheng Jing at the Chinese Institute of Psychology, examined how memory for words would be affected by the type of pictorial background they appeared on. Chinese and American college students and elderly people were asked to look at a large number of words. Some words were presented on a "social" background consisting of pictures of people, some on a background consisting of "nonsocial" objects such as flowers, and some on no background at all. After seeing the set of pictures, participants reported all the words they could recall. There was no difference between Chinese and Americans in recall of words initially presented on nonsocial backgrounds or on no background, but Chinese participants recalled more words that had been presented on social backgrounds than did American participants. Memory for the pictures of people apparently served as a *retrieval cue* for the words emblazoned on them, indicating that the Chinese had paid more attention to the social cues than the Americans.

There is good reason to believe that Westerners and Asians literally experience the world in very different ways. Westerners are the protagonists of their autobiographical novels; Asians are merely cast members in movies touching on their existences. Developmental psychologists Jessica Han, Michelle Leichtman, and Qi Wang asked four- and six-year-old American and Chinese children to report on daily events, such as the things they did at bedtime the night before or how they spent their last birthday. They found three remarkable things. First, although all children made more references to themselves than to others, the proportion of self-references was more than three times higher for American children than for Chinese children. Second, the Chinese children provided many small details about events and described them in a brief, matter-of-fact fashion. American children talked in a more leisurely way about many fewer events that were of personal interest to them. Third, American children made twice as many references to their own internal states, such as preferences and emotions, as did the Chinese children. In short, for American kids: "Well, enough about you; let's talk about me."

That Asians have a more holistic view of events, taking into perspective the orientation of other people, is also indicated by a study by social psychologists Dov Cohen and Alex Gunz. They asked North American students (mostly Canadian) and Asian students (a potpourri of students from Hong Kong, China, Taiwan, Korea, and various South and Southeast Asian countries) to recall specific instances of ten different situations in which they were the center of attention: for example, "being embarrassed." North Americans were more likely than Asians to reproduce the scene from their original point of view, looking outward. Asians were more likely to imagine the scene as an observer might, describing it from a third-person perspective.

It should be noted that for the studies described in this section, and for all studies conducted by our research teams in which some participants were tested in English and some in another language, we used the method of "back-translation" to ensure comparability. Materials were composed in language A and translated into language B. A native speaker of language B then translated the materials back into language A. If the native speaker of language A judged that the original and the back-translated version were identical in meaning, the materials were used as constructed. If not, the procedure was repeated.

My new Japanese student, Taka Masuda, was six feet two inches tall and weighed 220 pounds. He was a football player (yes, football—it's the third most popular sport in Japan). Needless to say, he was excited about going to his first Big Ten football game shortly after arriving at Michigan in the fall. He was in fact thrilled by the game, but he was appalled by the behavior of his fellow students. They kept standing up and blocking his view. In Japan, he told me, everyone learns from an early age to "watch your back." Nothing to do with paranoia—on the contrary, the point is to make sure that what you do doesn't impinge on the pleasure or convenience of others. The American students' indifference to the people behind him seemed unfathomably rude to him.

The behavior of American football fans motivated Masuda to test the hypothesis that Asians view the world through a wide-angle lens, whereas Westerners have tunnel vision. He achieved this by using a deceptively simple procedure. He showed eight color animated underwater vignettes, like the one reproduced in black-and-white at the top of the illustration on page 91, to students at Kyoto University and the University of Michigan. The scenes were all characterized by having one or more "focal" fish, which were larger, brighter, and faster-moving than anything else in the picture. Each scene also contained less rapidly moving animals, as well as plants, rocks, bubbles, etc. The scenes lasted about twenty seconds and were shown twice. After the second showing, participants were asked to say what they had seen. Their answers were coded as to what they referred to: focal fish, other active objects, background and inert objects, etc.

Americans and Japanese made about an equal number of references to the focal fish, but the Japanese made more than 60 percent more references to background elements, including the water, rocks, bubbles, and inert plants and animals. In addition, whereas Japanese and American participants made about equal numbers of references to movement involving active animals, the Japanese participants made almost twice as many references to relationships involving inert, background objects. Perhaps most tellingly, the very first sentence from the Japanese participants was likely to be one referring to the environment ("It looked like a pond"), whereas the first sentence from Americans was three times as likely to be one referring to the focal fish ("There was a big fish, maybe a trout, moving off to the left").

After participants had reported what they had seen in each vignette, they were shown still pictures of ninety-six objects, half of which they had seen before and half of which they hadn't. Their job was to say whether they had seen the objects before. Some of the objects that had actually been seen before were shown in their original environment and some were shown in a novel environment. Examples of both types are shown at the bottom of the illustration. The ability of the Japanese to recognize that

Recall Task

Recognition Task

Fish with Original
Background

Fish with Novel
Background

Examples of underwater scenes. *Top:* frame from film for recall task.
Bottom: still photos for recognition task.

they had seen an object before was substantially greater when the object was shown in the original environment than when it was shown in a new environment, suggesting that the object had become "bound" to the environment when seen initially and remained that way in memory. It made precisely no difference at all to Americans whether they saw the object in its initial environment or in a novel environment, suggesting that the perception of the object was fully separated from its environment.

In a follow-up study, Masuda and I showed various kinds of animals in different contexts to Americans and Japanese, this time measuring not only accuracy of recognition but also speed of processing. Again, the Japanese were more affected by the manipulation of background than were the Americans, making many more errors when the object was presented against a novel background than when it was presented against its original background. Moreover, the speed of Japanese judgments was impaired when the objects were presented against a novel background, whereas Americans' judgment speed was not affected.

Suppose you were approached by a man on the street who asked for directions. As you are talking to the person, two men come between you carrying a large sheet of plywood. The man who was talking to you grabs the end of the ply-wood and his confederate remains after the plywood procession is gone—as if he were the person you had been talking to originally. How likely is it, do you suppose, that you would fail to notice that you were talking to a changeling? Short of the two men being identical twins, you might guess that there is no chance of such an error.

In fact, it is easy to fool people with this trick. And in general people are remarkably impervious to the fact that some scene they are viewing has been altered in a substantial way. Film editors depend on this susceptibility: actors are standing in a slightly different relation to one another in a particular scene than they were at what is supposed to be a split second before; the cigarette is burned farther down earlier in the scene than later, and so on.

One implication of the notion that Easterners pay relatively more attention to the field than do Westerners is that we would expect the latter to be relatively blind to changes in objects in the background and to changes in relationships between objects. We might also expect that Westerners would be quicker to grasp alterations in salient foreground objects than Easterners would be. In order to examine this possibility, Masuda and I showed brief computer-generated color film clips to Japanese and American participants. The clips were almost, but not quite, identical. The illustration on page 94 shows black-and-white versions of one of the pairs. The scenes shown are frames from partway through the clips. The participant's job was to report in what way the clips differed. It can be seen that they differed in several respects. For example, the helicopter at the bottom has the black rotor to the left in one version and to the right in the other. The Concorde that is taking off has its landing gear down in one version and up in the other. Relationships between objects also differ. For example, the helicopter and the single-engine plane are closer together in one version than in the other. Finally, background details are different: The control tower has a different shape in one version than in the other.

Frame from Airport Site Movie: Version 1

Frame from Airport Site Movie: Version 2

Two versions of airport site movie.

As we anticipated, the Japanese participants noticed many more background differences between the two clips and many more relationship differences than Americans did. Americans were more likely to pick up changes in focal, foreground objects.

If Asians pay more attention to the environment than Westerners, we might expect that they would be more accurate in perceiving relationships between events. Exploring this question, Li-jun Ji, Kaiping Peng, and I presented Chinese and American participants with a split computer screen. On the left side of the screen we flashed one of two arbitrary figures: for example, a schematic medal or a schematic lightbulb. Immediately after, on the right side of the screen, we flashed one of another two arbitrary figures: for example, a pointing finger or a schematic coin. For some of the trials, there was no association whatever between what came up on the left and what came up on the right. For example, the coin was no more likely to come up on the right if it had been the medal that had come up on the left than if it had been the lightbulb on the left. For other trials, there was an association, sometimes a fairly strong one. We asked participants how strong they thought the association was on each set of trials and how confident they were that they were right.

Chinese participants reported stronger associations between what came up on the left and what came up on the right than did Americans, their confidence in their judgments was greater, and their confidence was better calibrated with the actual degree of association than was the case for Americans. Most strikingly, Americans showed the usual tendency found in covariation-detection studies of being overly influenced in their judgments by the first pairings seen. For example, if the lightbulb was frequently paired with the medal on early trials, the Americans tended to report that that had been the rule in general— even when that was not the case. The Chinese participants were subject to no such error.

Ji, Peng, and I also examined whether Americans are more capable of separating an object from its context than Asians. (There should be *some* advantage to the analytic, tunnel-vision perceptual style!) We presented East Asians (mostly Chinese and Koreans) and Americans with the Rod and Frame Test for "field dependence" invented by Witkin and his colleagues. In this test, you present participants with a long box, at the end of which is a rod. The rod can be manipulated independently of the box, which serves to frame the rod. The participant's task is to judge when the rod is exactly vertical, but the position of the frame inevitably influences judgments about the rod to a degree. People are deemed "field dependent" to the extent that their judgments about the verticality of the rod are affected by the context, that is, the orientation of the frame. We anticipated that the Asians would be more field dependent and indeed they were. They found it more difficult than did the Americans to make judgments about the position of the rod without being biased by the orientation of the frame.

CONTROLLING THE WORLD

If life is simple and you only have to keep your eye on the ball in order to achieve something, life is controllable. If life is complex and subject to changes of fortune without notice, it may not matter where the ball is; life is simply not easily controlled. Surveys show that Asians feel themselves to be in less control than their Western counterparts. And rather than attempting to control situations, they are likely to try to adjust to them. Social psychologists Beth Morling, Shinobu Kitayama, and Yuri Miyamoto asked Japanese and American students to tell them about incidents in their lives in which they had adjusted to some situation and incidents in which they had been in control of the situation. The incidents of adjustment were apparently more common for the Japanese, since the ones they remembered were on average more recent than was the case for Americans. Incidents of control were apparently more common for Americans than for Japanese because remembered control incidents were more recent for the Americans. Morling also asked her participants how they felt in each type of situation. The Americans, but not the Japanese, felt awkward, anxious, and incompetent when they had to adjust to a situation.

Other evidence also suggests that feeling in control is not as important for Asians as it is for Westerners. A survey of Asians, Asian Americans, and European Americans found that feeling in control of their lives was strongly associated with mental health for European Americans, but much less so for Asians and Asian Americans. In addition, feelings of well-being were enhanced more for Asians than for Americans by having other people around who might aid in providing control. And whereas Westerners seem to believe it's crucial for them to have direct, personal control, Asians seem to believe outcomes will be better for them if they are simply in the same boat with others.

Organizational psychologist P. Christopher Earley asked Chinese and American managers to work on managerial tasks under several different conditions. The managers thought they were either working alone; working with other members of their own group, that is, people from the same region of their country having interests similar to theirs; or working with members of an out-group, that is, people from another region of their country with whom they would have little if anything in common. The situation had been rigged so that the managers were really working alone in all conditions. In the "in-group" and "out-group" conditions, participants thought their performances would be assessed only at the group level and not at the individual level. Chinese managers performed better when they thought they were working with in-group members than when they thought they were alone or working with out-group members. Americans worked best when they thought they were alone, and it made no difference whether they thought they were working with an in-group or with an out-group.

The adage that "there's safety in numbers" may be Western in origin, but social psychologist Susumu Yamaguchi and his colleagues have shown that Japanese college students hold more closely to this tenet than do American students. They told participants in their study that they were interested in finding out the effects of an "unpleasant experience," namely swallowing a bitter drink, on performance of a particular task. Participants would be assigned either to a control condition or to the unpleasant experience condition. Just which condition would depend on the result of a lottery.

There were indeed two conditions in the experiment, but they were an "alone" condition and a "group" condition. Participants in the alone condition were told that they would draw four lottery tickets, each having a one-digit number on it. In the group condition, all participants believed they were part of a four-person group (whose members they never actually saw) and that each person would draw a lottery ticket. To participants in both conditions it was explained that the sum of the numbers on the four tickets would determine who would have to take the bitter drink. Yamaguchi and his colleagues asked participants how likely it was that they would be among the unlucky ones. (There was no objective reason for participants in either condition to think that the chances were any different in the alone condition than in the group condition.) The Japanese thought they were more likely to escape the unpleasant experience in the group condition. American men thought they were more likely to escape in the alone condition. American women behaved like Japanese, thinking escape was more likely if they were in a group.

The Yamaguchi study, as well as one described later in this section, is one of the rare studies finding that Western males and females differ from one another more than Eastern males and females do. In general, we either find gender differences for both Western and Eastern cultures—of about the same magnitude—or we find gender differences for neither culture. As would be expected, given our theory about the social origins of the cognitive and perceptual differences, females of both cultures tend to be more holistic in their orientation than males, but we find this only about half the time, and the gender differences are always smaller than the cultural differences. We have been unable to characterize the difference between tasks for which we find gender differences and those for which we don't.

Thus, to the Asian, the world is a complex place, composed of continuous substances, understandable in terms of the whole rather than in terms of the parts, and subject more to collective than to personal control. To the Westerner, the world is a relatively simple place, composed of discrete objects that can be understood without undue attention to context, and highly subject to personal control. Very different worlds indeed.

The world of Westerners, however, is not as controllable as they think. Ellen Langer, a social psychologist, identified a foible she called the "illusion of control," which she defined as an expectation that personal success is greater than the objective probability would warrant. The illusion can sometimes be a helpful thing. In one study, for example, people have been found to perform better on routine tasks when they believe mistakenly that they can control a loud, distracting noise that occurred periodically during the tasks. On the other hand, there are also some demonstrations of the illusion that make us look pretty silly. In my favorite study, Langer approached people in an office building and asked whether they would like to buy a lottery ticket for a dollar. If the person said yes, she then either handed the person a lottery ticket or fanned out a bunch of them and asked the person to choose one. Two weeks later, she approached all those who had bought a ticket, saying that lots of people wanted to buy a ticket, but there were none left. Would the person be willing to sell the ticket back, and if so, what would the price be? On average, the people she had handed the ticket to were willing to sell the ticket back for about two dollars, but the people who had been allowed to choose their tickets held out for almost nine!

Much of what we know implies that Asians would be less susceptible to such illusions of control than Westerners, as well as less concerned about issues of control altogether. Ji, Peng, and I tested these notions with new versions of our covariation detection test and the Rod and Frame Test.

In a twist on the covariation detection task, in which the goal was to determine how likely it was that one particular object would appear on the right side of a computer screen given that another particular object had appeared on the left, we gave the participants control over which object would be presented on the left of the computer screen and allowed them to choose how much time would elapse on each trial between presentation of the object on the left and presentation of the object on the right. Under these circumstances, the Americans saw as much covariation as the Chinese did and they were as confident as the Chinese. Moreover, the Americans were reasonably accurate about the degree of covariation they saw, whereas the Chinese were actually slightly less accurate when they had control than when they didn't.

In a variation of the Rod and Frame Test, we gave the

participants control of the rod, allowing them to rotate it themselves. Under these circumstances, Americans became more confident about the accuracy of their judgments, whereas East Asians did not become more confident. And American men, who were the most accurate of the groups to begin with, actually became more accurate still. Accuracy for East Asians and for American women was unaffected by being given control.

STABILITY OR CHANGE?

> When we think about the future of the world,
> we always have in mind its being where it
> would be if it continued to move as we see it
> moving now. We do not realize that it moves
> not in a straight line . . . and that its direction
> changes constantly.
> —PHILOSOPHER LUDWIG WITTGENSTEIN

> [We tend] to postulate that tomorrow will be
> the same as today; likewise, when we are aware
> of movement, we assume that tomorrow will
> differ from today in the same way as today differs from yesterday. . . . The lifespan of man has
> become longer; it will become longer still. The
> number of work hours in the year has decreased;
> it will decrease yet further. . . . The sharper our
> awareness of a past movement, the stronger our
> conviction of its future continuation.
> —POLITICAL PHILOSOPHER BERTRAND DE JOUVENAL

As it turns out, "our" is rather too strong a generalization. Ancient Greek philosophers were powerfully inclined to believe that things don't change much or, if they really are changing, future change will continue in the same direction, and at the same rate, as current change. And the same is true for ordinary modern Westerners. But like ancient Taoists and Confucian philosophers, ordinary modern Asians believe that things are constantly changing; and movement in a particular direction, far from indicating future changes in the same direction, may be a sign that events are about to reverse direction.

These differing assumptions about change can be derived from different understandings about the complexity of the world, which in turn are a consequence of attending to a small part of the environment versus a lot of it. If the world appears a simple place because we're not paying attention to much of it, then not much change is to be expected. If change is occurring, then there is no reason to assume that it will do anything but continue in the same direction. But if the world seems to be a highly complicated place because we're noticing so much, then stability will be the exception and change will be the rule. The greater the number of factors operating, the greater the likelihood that some variable will alter the rate of change or even reverse its direction. The specifically cyclical assumptions of the Tao may spring from these theories about complexity. Or it could be the other way around: The belief that the world is constantly reverting to prior states may prompt the assumption of complexity. To be dialectical about it, probably both trends are operative, and feed each other . . . in a cycle!

With Li-jun Ji, a student at Michigan at the time and Yanji Su, a colleague at Beijing University, I studied Chinese and American beliefs about change. In one study, we asked University of Michigan students and Beijing University students how likely they thought it was that some state of affairs would undergo a radical change. For example: "Lucia and Jeff are both seniors at the same university. They have been dating each other for two years. How likely is it that they will break up after graduation?"

There were four such items asking about the probability of change. In all four instances, the Chinese regarded change as more likely than did the Americans. On average, Chinese thought change was likely about 50 percent of the time and Americans thought change was likely about 30 percent of the time.

In a second study, Ji, Su, and I showed Beijing and Michigan participants twelve graphs in a booklet. Each graph showed an alleged trend charted over time, such as world economy growth rate or world cancer death rate. For example: The global economy growth rates (annual percentage change in real GDP) were 3.2 percent, 2.8 percent, and 2.0 percent for 1995, 1997, and 1999 respectively.

We asked the participants how likely they thought it was that the global economic growth rate would go up, go down, or remain the same for 2001.

The trends we presented were either growing or declining and the rate of change was either accelerating or decelerating. The illustration shows a positively accelerated growth curve and a negatively accelerated growth curve. We reasoned that the greater the increase in the

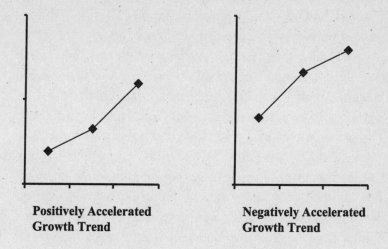

**Positivey Accelerated
Growth Trend** **Negatively Accelerated
Growth Trend**

Examples of positively and negatively accelerated trends.

rate of change, the more likely it was that the Chinese would anticipate slowing or even reversal of the trend; as more rapid change in a given direction should signal reversal in the near future. For Americans, however, increases in acceleration might be a particularly strong indicator of continued movement in a particular direction. So we expected differences between Chinese and Americans to be greater when assessing positively accelerated trends than when assessing negatively accelerated trends.

We found that, as expected, Americans made more predictions consistent with the trends we showed them than did Chinese. In fact, this was true for all twelve graphs we showed. If a particular trend went up, the Americans were more likely to predict that it would continue going up than were the Chinese. If the trend went down, the Americans were more likely to predict decline would continue than were the Chinese. And these differences were, as anticipated, greater for the positively accelerated trends than for the negatively accelerated ones.

In a variant of this study, we showed the same set of twelve graphs with their three initial data points to a new group of participants and asked them to actually plot what they thought the next two data points might be. Americans were likely to continue the trend in the same direction, and at the same rate, as could be extrapolated from the previous points. The Chinese on average predicted a leveling off of change and were several times more likely to predict a reversal in direction of change than Americans were. Again, these trends were more marked when graphs were positively accelerated than when they were negatively accelerated.

Beliefs about linear versus cyclical movement apply to change over very great time spans. Thomas More's 1516 political essay speculated on the form of perfect government. More invented the term "Utopia" to name his society. The word is a pun on a Greek root meaning both "nowhere" and "good place." More's Utopia was scarcely the first and certainly not the last in a long line of Western creations, including Plato's Republic, Puritanism, Shaker communities, Mormonism, the American and French revolutions, communism, and fascism. With the chief exceptions of Utopias modeled on the biblical ideas of the Garden of Eden and the promise of the New Jerusalem, Western Utopias have generally had five salient characteristics—all of which make them vastly different from the conviction of Confucius and other early Chinese thinkers that the perfect world existed in the past and that we could hope only to strive to move from our current low estate back to that time of perfection.

In Western Utopias:

- there is steady, more or less linear progress toward them;
- once attained, they become a permanent state;
- they are reached through human effort rather than Fate or divine intervention;
- they are usually egalitarian; and
- they are usually based on a few extreme assumptions about human nature.

These attributes are in many ways the very antithesis of the future as it might be conceived by the Eastern mind, which is inclined to find the Middle Way between extremes and assumes reversion rather than advance.

It is worth noting here that the ancient Hebrews were in these respects closer to the Chinese than to the Greeks. Their Utopia—the Garden of Eden—was in the past and they hoped at most for a restoration. Their notion of the nature of change was similar to that of the Chinese—they had a clear notion of the yin and yang of life. Hebrew prophets of the eighth century B.C. sold real estate when things were going well for the Jews—because they felt sure that things would soon take a turn for the worse—and bought when things were going badly! This attitude toward life survives in the modern Jewish community, as is conveyed by countless jokes. Son: "Mom, guess what—I won a Pontiac in the raffle!" Mom: "Oy, the taxes alone will put us in the poorhouse."

If the differences in assumptions about the direction of human progress persist, and if people make analogies to the direction of a single human life, we might find that Westerners believe that their own futures will move continuously in a single direction—from bad to good or good to bad. East Asians might expect their lives to undergo reversals of fortune—from good to bad to good, or from bad to good to bad. In order to examine these possibilities, Ji, Su, and I asked college students at Michigan and Beijing to predict the course of their own life happiness. We showed them eighteen different trends to choose from. Six were linear—straight up or down but with oscillations along the way. Twelve were nonlinear—either stopping or reversing the initial direction of life change. Almost half of the Americans chose one of the six linear life courses as the most probable, whereas fewer than a third of the Chinese choices were linear. (Choices were not due to either group having uniformly optimistic or pessimistic assumptions about life course. The two groups were equally likely to feel they would end up happy and equally likely to feel they would end up unhappy.)

Like their ancient predecessors, then, East Asians believe that the world is full of change and that what goes around comes around. Westerners (or at any rate, Americans—we have no data on other Westerners at this point) appear to believe that what goes up needn't come down.

In chapter 3, we saw that the social organization and practices of modern Asians resemble those of the ancient Chinese and the social organization and practices of modern Europeans resemble those of the ancient Greeks. In this chapter we've seen that modern Asians, like the ancient Chinese, view the world in holistic terms: They see a great deal of the field, especially background events; they are skilled in observing relationships between events; they regard the world as complex and highly changeable and its components as interrelated; they see events as moving in cycles between extremes; and they feel that control over events requires coordination with others. Modern Westerners, like the ancient Greeks, see the world in analytic, atomistic terms; they see objects as discrete and separate from their environments; they see events as moving in linear fashion when they move at all; and they feel themselves to be personally in control of events even when they are not. Not only are worldviews different in a conceptual way, but also the world is literally *viewed* in different ways. Asians see the big picture and they see objects in relation to their environments—so much so that it can be difficult for them to visually separate objects from their environments. Westerners focus on objects while slighting the field and they literally see fewer objects and relationships in the environment than do Asians.

If some people view the world through a wide-angle lens and see objects in contexts, whereas others focus primarily on the object and its properties, then it seems likely that the two sorts of people will explain events quite differently. People having a wide-angle view might be inclined to see events as being caused by complex, interrelated contextual factors whereas people having a relatively narrow focus might be prone to explain events primarily in terms of properties of objects. In the next chapter, we'll see whether the different worldviews are indeed associated with different kinds of causal explanations for the same event.

CHAPTER 5

"THE BAD SEED" OR "THE OTHER BOYS MADE HIM DO IT"?

In 1991 a Chinese physics student at the University of Iowa named Gang Lu lost an award competition. He appealed the decision unsuccessfully and he subsequently failed to obtain an academic job. On October 31, he entered the physics department and shot his adviser, the person who had handled his appeal, several fellow students and bystanders, and then himself.

Michael Moris, a graduate student at Michigan at the time, noticed that the explanations for Gang Lu's behavior in the campus newspapers focused almost entirely on Lu's presumed qualities—the murderer's psychological foibles ("very bad temper," "sinister edge to his character"), attitudes ("personal belief that guns were an important means to redress grievances"), and psychological problems ("a darkly disturbed man who drove himself to success and destruction," "a psychological problem with being challenged"). He asked his fellow student Kaiping Peng what kinds of accounts of the murder were being given in Chinese newpapers. They could scarcely have been more different. Chinese reporters emphasized causes that had to do with the context in which Lu operated. Explanations centered on Lu's relationships ("did not get along with his adviser," "rivalry with slain student," "isolation from Chinese community"), pressures in Chinese society ("victim of Chinese 'Top Student' educational policy") and aspects of the American context ("availability of guns in the U.S.").

In order to be sure that their impressions were accurate, Morris and Peng carried out a systematic content analysis of reports in the *New York Times* and the Chinese-language newspaper the *World Journal*. This objective procedure showed that their initial observations were correct. Should the different causal attributions be regarded as mere chauvinism? The American reporters blamed the perpetrator, who happened to be Chinese, whereas the Chinese reporters, perhaps protecting one of their own, blamed situational factors. As it happens, a "control" mass murder allows us to see whether it was chauvinism or worldview that produced the differences in explanation patterns.

In the same year that Gang Lu committed his murders and suicide, an American postal worker in Royal Oak, Michigan, named Thomas McIlvane lost his job. He appealed the decision unsuccessfully to his union and subsequently failed to find a full-time replacement job. On

November 14, he entered the post office where he had previously worked and shot his supervisor, the person who handled his appeal, several fellow workers and bystanders, and then himself.

Morris and Peng performed the same kind of content analysis on the *New York Times* and *World Journal* reports of the McIlvane mass murder that they did for the Gang Lu mass murder. They found exactly the same trends as for the Chinese murderer. American reporters focused on McIlvane's personal dispositions—attitudes and traits inferred from past behavior ("repeatedly threatened violence," "had a short fuse," "was a martial arts enthusiast," "mentally unstable"). Chinese reporters emphasized situational factors influencing McIlvane ("gunman had been recently fired," "post office supervisor was his enemy," "influenced by example of a recent mass slaying in Texas").

Morris and Peng gave descriptions of the murders to American and Chinese college students and asked them to rate the importance of a large number of presumed personal attributes and situational factors culled from the newspaper reports. American students, whether explaining the American mass murder or the Chinese one, placed more emphasis on the murderer's presumed dispositions. Chinese students stressed situational factors for both mass murders. Even more impressively, Morris and Peng listed a number of situational factors and asked participants to judge whether, if circumstances had been different, the murder might not have occurred. They asked, for example, if the tragedies might have been averted "if Lu had received a job" or "if McIlvane had had many friends or relatives in Royal Oak." Americans and Chinese participants responded very differently. The Chinese thought that, in many cases, the murders might very well not have occurred. But the Americans, because of their conviction that it was the murderer's long-established dispositions that were the key to his rampage, felt it was likely that the killings would have occurred regardless of whether circumstances had been different.

CAUSAL ATTRIBUTION EAST AND WEST

It should come as no surprise that Chinese people are inclined to attribute behavior to context and Americans tend to attribute the same behavior to the actor. We saw in the last chapter that East Asians attend more to context than do Americans. And what captures one's attention is what one is likely to regard as causally important. The converse seems equally plausible: If one thinks something is causally important one is likely to attend to it. So a cycle gets established whereby theories about causality and focus of attention reinforce each other.

There is ample evidence that the causal attribution differences mirror the attention differences. The first cross-cultural study of causal attribution, by developmental psychologist Joan Miller, compared Hindu East Indians and Americans. She asked her middle-aged, middle-class participants to describe behavior of an acquaintance that they "considered a wrong thing to have done" and behavior on the part of an acquaintance that they "considered good for someone else." She then asked her participants to explain why the people behaved as they did. American participants tended to explain the behavior in terms of presumed personality traits and other dispositions of the actor: "Sally is considerate, outgoing, and friendly." The Americans made twice as many such attributions as the Indians. Indians tended to explain behavior in terms of contextual factors: "It was dark and there was no one else to help." The Indians gave twice as many contextual explanations as Americans did.

Americans and Indians didn't give different sorts of answers because they had somehow described different kinds of events. When Miller asked Americans to explain behaviors mentioned by Indians, Americans explained them using the same sorts of dispositional explanations they gave for the behaviors they generated themselves. In a particularly important additional demonstration, Miller showed that it takes time to learn how to explain behavior in the culturally sanctioned way. Children in the two cultures didn't differ in the sorts of explanations they gave. Not until adolescence did Indians and Americans begin to diverge in their explanations. To put the icing on the cake of this elegant study, Miller also questioned Anglo-Indians, whose culture is Westernized to a degree. Their attributions, both for dispositions and for contexts, were midway between those of Hindu Indians and Americans.

A favorite activity around the water cooler of a Monday morning is discussing why the game was won or lost. It turns out that the reasons people give for victory or defeat are different in America and Asia. Organizational psychologist Fiona Lee and her colleagues analyzed what sportswriters reported about the causal attributions of soccer coaches and players in the U.S. and Hong Kong. Americans saw outcomes as being due mostly to the abilities of individual players: "Freshman Simpson leads the team in scoring with eleven goals, but its success lies in its defense." "We've got a very good keeper in Bo Oshoniyi, who was defensive MVP of the finals last year . . ." The attributions of Hong Kong athletes and coaches were more likely to refer to the other team and the context: "We were lucky to go in at the interval with a one-goal advantage and I was always confident we could hold them off. I guess South China was a bit tired after having played in a quadrangular tournament in China."

Attributional differences between Asians and Westerners go deeper than accounts of human behavior. Morris and Peng showed that Chinese tend to attribute the behavior of fish shown in video scenes to external factors and Americans to internal factors. Peng and his colleagues have shown that the differences between Easterners and Westerners go deeper still—to the perception of physical causality. They showed abstract cartoons, like those illustrated on page 117, to Chinese and American women. Each cartoon showed movement of some kind that could be interpreted as hydraulic, magnetic, or aerodynamic. As intended, participants interpreted the top sequence in the illustration as a light object (a "ball") coming to "float" on the liquid. In the cartoon based on the picture beneath it, the circle dropped past the upper line and came to rest on the lower line. As intended, participants saw this movement as a heavy object dropping to the bottom of a container of liquid. Participants were asked to what extent they thought that the object's movements seemed influenced by internal factors (something inside the object or belonging to it had caused it to drop). The Americans reported that they thought the movements were caused more by internal factors than did the Chinese.

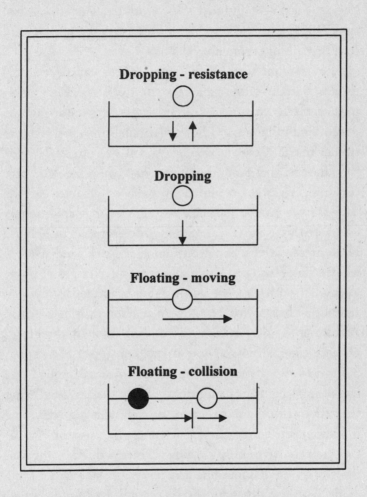

Trajectories of motion in computer displays
suggestive of liquid in a container.

◌

The British were in charge of Hong Kong for one hundred years and the children there learn English no later than elementary school. Western influence, both culturally and linguistically, remains strong, even under Chinese control since 1997. This makes the city an interesting laboratory for purposes of cross-cultural study.

It turns out that Hong Kong citizens can be encouraged to think in either an Eastern or a Western way by presenting them with images that suggest one culture or the other. Ying-yi Hong and her colleagues showed a vignette similar to the Morris and Peng fish cartoons to students at the University of Hong Kong. But first, they showed them pictures suggestive of either Western or Eastern culture. They showed some participants pictures that are strongly associated with American culture: for example, the House of Representatives, a cowboy on horseback, and Mickey Mouse. They showed other participants pictures strongly associated with Chinese culture: for example, a dragon, a temple, and men writing Chinese characters using a brush. A third group of participants were shown neutral pictures of landscapes. After showing participants a set of pictures, Hong and her colleagues showed them a cartoon of one fish swimming in front of other fish and asked them what they thought was the major reason for the fish's swimming in front of the other fish. Participants who saw the American pictures gave more reasons having to do with motivations of the individual fish and fewer explanations having to do with the other fish or the context than did participants who saw the Chinese pictures. Participants who saw the neutral pictures were in the middle.

Peng and his colleague Eric Knowles studied Asian Americans and found that they could "prime" either their participants' Asian selves or their American selves. They showed students a battery of vignettes of physical movement like those portrayed in the illustration, and asked them to rate the extent to which the object's movement was due to dispositional factors (e.g., shape, weight) vs. contextual factors (e.g., gravity, friction). But first they asked participants either to recall an experience they had that made identity as an American apparent to them or to recall an experience that made their Asian identity apparent. The primes had an effect. Participants who had their American identity primed rated causes internal to the object to be more important than did participants who had their Asian identity primed.

Ara Norenzayan, Incheol Choi, and I asked Korean and American college students a number of questions intended to plumb their theories about the causes of behavior. We asked them to rate the degree to which several paragraphs captured their views about the reasons people behave as they do. The first couple of sentences of each paragraph are reproduced below.

1) How people behave is mostly determined by their personality. One's personality predisposes and guides an individual to behave in one way, not in another way, no matter what circumstances the person is in.

2) How people behave is mostly determined by the situation in which they find themselves. Situa-

tional power is so strong that we can say it has
more influence on behavior than one's personality.
3) How people behave is always jointly determined
by their personality and the situation in which
they find themselves. We cannot claim that either
personality or the situation is the only determi-
nant of our behavior.

Koreans and Americans regarded personality (1) as
equally important in determining behavior, but Koreans
rated situational factors (2) and the interaction between
situations and personalities (3) as more important than
Americans did.

We also asked participants several questions about
their beliefs regarding the malleability of personality. For
example, we asked whether they thought that someone's
personality is something about them that they can't
change very much. The Koreans thought that personalities
are more subject to change than the Americans did.

It should hardly be surprising that Americans regard
personalities as relatively fixed and Asians regard them as
more malleable. This is consistent with the long Western
tradition of regarding the world as being largely static and
the long Eastern tradition of viewing the world as constantly changing.

Social psychologists Michael Morris, Kwok Leung, and
Sheena Sethi (Iyengar) have shown that Easterners and
Westerners have preferences for different kinds of negotiation strategies, which may be related to views about plia-
bility of character. Hong Kong and American participants
were asked what kind of adjudication they would prefer

when they had to come to an agreement with someone who had behaved in ways that could be construed as belligerent and unreasonable. Hong Kong participants preferred inquisitorial adjudication in which a third party questions the disputants and tries to make a mutually agreeable judgment, whereas Americans were more likely to prefer adversarial adjudication with representation by lawyers.

Should we assume that Asians have theories of human personality that are fundamentally different from those of Westerners? Do Asians believe that people differ from one another only very slightly? Or do they see differences, but in terms of traits that would seem odd or irrelevant to Westerners?

Probably the answer to all of these questions is no. When I was in China in 1982, toward the end of the Cultural Revolution, the society was still somewhat shell-shocked and secretive, having just spent thirty years undergoing a convulsive social and economic experiment. The culture was and is dramatically different from that of the West in ways that I could not have articulated at the time. As this book shows, there are marked differences in worldviews, perception, and thought processes. Yet within three weeks I found that I was able to gossip with my hosts about other Chinese. We could talk about Fung's decency and humility, Chan's arrogance, Lin's reserve, understanding each other perfectly. Fortunately there is better evidence than my anecdote available. Researchers have produced a large amount of evidence indicating that theories of personality in the East are quite similar to those in the West. Major personality trait factors—labeled the Big Five by personality theorists—are repeatedly found in Western populations. These same factors tend to be found when the Western personality tests are translated and given to Chinese, Koreans, or Japanese, though sometimes only four of the factors are identified.

Cultural psychologists Kuo-shu Yang and Michael Bond have found that there is also pretty good replication when test items are not translated from Western languages, but rather are generated by researchers using items based on behavior descriptions common in the local culture. In a subsequent effort to develop an "indigenous" Chinese personality inventory, Fanny Cheung and her colleagues selected items descriptive of personality from popular contemporary Chinese novels, books of Chinese proverbs, and descriptions of themselves and others offered by ordinary people and by professional psychologists. Based on these items, Cheung and her colleagues constructed a "Chinese Personality Assessment Inventory." They administered this inventory to a large sample of people in Hong Kong and mainland China. They found four factors, three of which corresponded roughly to extraversion, neuroticism, and conscientiousness, the most robust of the Big Five factors in the West. Interestingly, the researchers found a factor that does not emerge in Western-developed tests, which they described as the "Chinese tradition" factor, a construct that captures personality descriptions related to maintenance of interpersonal and inner harmony. It would be intriguing to see if this factor could be found in a version of the Chinese inventory if it were to be translated into Western languages. Harmony is not the first characteristic that occurs to Western researchers when thinking about personalities, but the dimension might nevertheless be meaningful to Westerners.

Avoiding the Fundamental Attribution Error

It appears that Easterners and Westerners don't seem to differ that much in the personality dimensions they use. Why is it then that Westerners rely so much more heavily on personality traits in explaining behavior? The answer seems to be that Easterners are more likely to notice important situational factors and to realize that they play a role in producing behavior. As a consequence, East Asians are less susceptible to what social psychologist Lee Ross labeled the "Fundamental Attribution Error" (or FAE for short).

Imagine that you see a college student being asked to show possible donors around the campus for a day and that for this service the student is offered only a small amount of money—less than the minimum wage—and imagine that the student refuses. Do you suppose you would think it is likely that the student would volunteer to help in an upcoming Red Cross blood drive? Probably not very likely. But suppose a friend of yours had seen another student offered a reasonable amount of money— say, 50 percent above the minimum wage—to show the donors around and the student had agreed to do so. Do you suppose the friend would think it is likely that the student would help in the blood drive? Probably more likely than you thought your student would be. If so, both you and your friend would be showing a version of the FAE: attributing behavior to a presumed disposition of the person rather than to an important situational factor—namely money—that was the primary driving force behind the behavior.

This error—ignoring the situation and inventing strong dispositional explanations for behavior—is a highly pervasive one. It makes people mistakenly confident that a person they see being interviewed for an important job is rather nervous by nature, that a person they see being withdrawn at a particular party (where the person happens to know no one) is rather shy in general, that a person who gives a good talk on a subject they know well, to a familiar audience, is a polished speaker and a confident person to boot.

The first solid experimental demonstration of the error was by the noted social psychologist Edward E. Jones and his colleagues. In a study published in 1967, they asked college students to read a speech or essay allegedly written by another student. This other student will be called the "target." It was made clear that the target had been required to write the speech or essay upholding a particular side of a particular issue. For example, the target had been told to write an essay in a political science class favoring Castro's Cuba or to give a speech in a debate class opposing the legalization of marijuana. Participants were asked to indicate what they thought was the actual opinion of the target student who wrote the essay or gave the speech. The sharp situational constraints should have made the participants recognize that they had learned nothing about the target's real views, but in fact they were heavily influenced by what the target said. If the target said he was in favor of Castro's handling of Cuba, participants assumed he was actually inclined toward that opinion; if the target said he was opposed to the legalization of marijuana, participants tended to assume he was of that view.

As it turns out, this illusion is sufficiently powerful that even East Asians are susceptible. Chinese, Japanese, and Koreans have all participated in versions of this experiment and have been found to infer that the targets actually have attitudes corresponding to the views they read in the essay. But there is a difference between East Asian and American susceptibility: East Asians do not make the error if they are first placed in the target's shoes. Incheol Choi and I placed participants themselves into the situation of being required to write an essay on a particular topic, taking a particular stance, and using a particular set of four arguments in writing their essay. Then they read an essay by a person who, they knew, had been in the same situation they themselves had been. This had precisely no effect on Americans: Their dispositional inferences about others were as strong as if they had not themselves experienced exactly the target person's situation. But the experience rendered Koreans almost impervious to the error.

Other evidence indicates that making situational factors salient has a greater effect on Asians than on Westerners. Ara Norenzayan, Incheol Choi, and I asked American and Korean college students to read one of two scenarios and then to guess whether a target person would give someone bus fare. Both scenarios began in the following way:

You just met a new neighbor, Jim. As you and Jim are taking a walk in the neighborhood, a well-dressed man approaches Jim and explains that his car is broken down and he needs to call a mechanic. Then with a somewhat embarrassed voice, the man asks Jim for a quarter to make the phone call. You find that Jim searches his pocket and, after finding a quarter, gives it to the man. On another day Jim is walking toward the bus stop to catch the bus to work. As he is walking, a teenager carrying some books approaches Jim and politely asks him if he can borrow a dollar for a bus ride, explaining that he forgot his wallet at home and needs to get a ride to school.

In a version of the scenario read by one group of participants, Jim searches his pocket and discovers that he has several dollars; in a version read by other participants he discovers that he has only enough money for his own bus fare. Korean participants were more likely to recognize that Jim would be inclined to give the teenager the money if he finds he has several dollars than if he finds he has only one.

We gave participants a total of six different scenarios, each having their two different versions, and found that for each one the Koreans were more responsive to the situational information than the Americans were, predicting that a given behavior was more likely if situational factors facilitated it than if situational factors discouraged it.

So the evidence on causal attribution dovetails with

the evidence on perception. Westerners attend primarily to the focal object or person and Asians attend more broadly to the field and to the relations between the object and the field. Westerners tend to assume that events are caused by the object and Asians are inclined to assign greater importance to the context.

Building Causal Models

Differences in causal reasoning between Easterners and Westerners are broader than just preferences for field vs. object. Westerners seem to engage in more causal attribution, period. Historian Masako Watanabe has made this point beautifully in her studies of the ways Japanese and American elementary school and college students and their teachers deal with historical events.

Japanese teachers begin with setting the context of a given set of events in some detail. They then proceed through the important events in chronological order, linking each event to its successor. Teachers encourage their students to imagine the mental and emotional states of historical figures by thinking about the analogy between their situations and situations of the students' everyday lives. The actions are then explained in terms of these feelings. Emphasis is put on the "initial" event that serves as the impetus to subsequent events. Students are regarded as having good ability to think historically when they show empathy with the historical figures, including those who were Japan's enemies. "How" questions are asked frequently—about twice as often as in American classrooms.

American teachers spend less time setting the context than Japanese teachers do. They begin with the *outcome*, rather than with the initial event or catalyst. The chronological order of events is destroyed in presentation. Instead, the presentation is dictated by discussion of the causal factors assumed to be important ("The Ottoman empire collapsed for three major reasons"). Students are considered to have good ability to reason historically when they are capable of adducing evidence to fit their causal model of the outcome. "Why" questions are asked twice as frequently in American classrooms as in Japanese classrooms.

Watanabe labels American historical analysis as "backward" reasoning because events are presented in effect-cause order. She notes the similarity of this to goal-oriented reasoning: define the goal to be achieved and develop a model that will allow you to attain it. She also notes that goal orientation is more characteristic of Westerners, with their sense of personal agency, than it is of Asians. This insight helps us to understand why it was the Greeks and not the Chinese who engaged in causal modeling of natural phenomena. Modeling events in a "backward," causal-analysis fashion comes more naturally to people who are at liberty to set their own goals with respect to an object and to come up with schemes to achieve them. Watanabe quotes an American instructor of English as a second language as saying that "it is very difficult for American teachers to understand Japanese students' essays because we don't see any causality in them, and . . . the relation of cause and effect is elementary logic in the United States."

Consistent with the lesser complexity of the world they live in, Westerners see fewer factors as being relevant to an understanding of the world than Easterners do. Incheol Choi and his colleagues described the Chinese physics student murder story to American and Korean participants. Choi and colleagues then provided one hundred items of information concerning the student, the professor, the school, and so on and asked their participants to rule out factors that could not be considered to be of possible relevance for establishing a motive for the slaying. Korean participants regarded only 37 percent of the items of information as irrelevant. American participants thought 55 percent of the items were likely to be irrelevant. (They also studied Asian American participants and found them to be in between European Americans and Koreans.)

Choi and his colleagues also found evidence that the tendency to see so many factors as relevant to the outcome was related to the degree to which the individual held holistic beliefs about the world. They asked their participants to answer a "holism" questionnaire indicating the extent to which they believed that events are related to one another. Some examples:

- Everything in the universe is somehow related to everything else.
- It's not possible to understand the pieces without considering the whole picture.

Choi and colleagues found that Koreans were more holistic in their beliefs than Americans. Moreover, the more holistic the individual, whether American or Korean, the more reluctant to assume that a particular item of information might be irrelevant.

But open-mindedness and the belief that the world is complex can also have their disadvantages, as we'll see next.

Avoiding Hindsight

The Soviet Union's demise in 1991 may be one of the few historical events that has not seemed inevitable after the fact to large numbers of historians, both lay and professional. The fall of the Roman Empire, the rise of the Third Reich, and the American success in reaching the moon before the Russians, not to mention less momentous events, are routinely seen as inevitable by commentators, who, one strongly suspects, could not have predicted them. We tend to have two problems when we try to "predict" the past: (1) believing that, at least in retrospect, it can be seen that events could not have turned out other than they did; and (2) even thinking that in fact one easily could have predicted *in advance* that events would have turned out as they did.

How do we know that people are inclined to make these errors? Cognitive psychologist Baruch Fischhoff worked out a clever method for showing that people overestimate the extent to which they could have predicted the outcome of a given event and are less surprised by unusual turns of events than they should be. Fischhoff gave his participants enough information to set the stage for various historical events. For example, Fischhoff described the situation in 1814 in Bengal when the British were attempting to consolidate their control of India. They had to deal with raids by the Gurkas of Nepal. The British commander decided to deal with the Gurkas by invading their mountain territory. Details of the situation at the time of the invasion were provided and Fischhoff then asked his participants how likely they thought various outcomes were. He gave other participants the same information, but also told them the actual outcome (a stalemate). He asked these participants how likely they *would have thought* the outcome would be if they had not been told what it was. Fischhoff found that if his participants knew the outcome, they routinely overestimated the likelihood they would have assigned to it in advance.

Incheol Choi and I reasoned that it may be easier to avoid the hindsight fallacy if one is inclined to construct explicit causal models of the world. Explicit models are likely to turn up factors that could suggest more than one outcome and as a result one may be less inclined to be confident that some particular outcome would have occurred. Moreover, one can be surprised when one's predictions turn out to be wrong. Surprise is likely to prompt a search for possibly relevant factors and to revision of the model that in turn can result in a more accurate understanding of the world. On the other hand, if modeling is less explicit, and if large numbers of factors are considered to be potentially relevant to any given outcome, then it may be easy to think of reasons why a particular event might have turned out the way it did. We tested these notions in a series of experiments comparing Koreans and Americans.

We told participants in one study about a young semi-nary student, who, they were assured, was a very kind and religious person. Heading across campus to deliver a ser-mon, he encountered a man lying in a doorway asking for help. We told participants that the seminarian was late to deliver his sermon.

In condition A, participants did not know what the seminary student had done, and we asked them to tell us what they thought was the probability that the target would help and how surprised they would be if they were to find out that he had not helped. Both Koreans and Americans reported about an 80 percent probability that the target would help and indicated that they would be quite surprised if he did not. In condition B, we told partic-ipants that the seminary student had helped the victim and in condition C, we told participants that the target had not helped the victim. Participants in conditions B and C were asked what they believed they *would have* regarded as the probability that the student would have helped—if in fact they had not been told what he did—and also how sur-prised they were by his actual behavior. Again, both Kore-ans and Americans in condition B indicated they would have thought the probability of helping was about 80 per-cent and both groups reported no surprise that he did help. Americans in condition C, in which the student unexpect-edly did not help the victim, also reported that they would have thought the probability was about 80 percent that the student would have helped and they reported a great deal of surprise that he did not do so. In contrast, Koreans in condition C reported that they would have thought the probability was only about 50 percent that the student would have helped and they reported little surprise that he did not. So Americans experienced surprise where Koreans did not and Koreans showed a pronounced hindsight bias, with many indicating they thought they knew something all along which in fact they did not. (The scenario in our study described an actual experiment done with students at Princeton Theological Seminary. The fine young men of that study were very likely to offer help to the groaning man in the doorway—unless they were in a hurry, in which case most did not.)

Choi and I conducted another study that indicates that Easterners are not as surprised by unanticipated outcomes as Americans are. We described studies to American and Korean participants and either gave them one hypothesis about each study or two conflicting hypotheses—one that predicted the actual outcome of the study and one that predicted the opposite. For example, some participants were told about a study examining the hypothesis that realism increases mental health. Other participants were told that that hypothesis was being considered, as well as an alternative one that optimism promotes mental health. Then all participants read that actual research findings indicate that realism promotes mental health. We asked participants to indicate how surprising and interesting the finding was. Americans reported being more surprised—and found the study to be more interesting —when we had presented two strongly competing hypotheses, whereas Koreans were no more surprised or interested when presented with two opposing hypotheses than when presented only with the one that predicted the actual finding.

ॐ

Easterners are almost surely closer to the truth than Westerners in their belief that the world is a highly complicated place and Westerners are undoubtedly often far too simple-minded in their explicit models of the world. Easterners' failure to be surprised as often as they should may be a small price to pay for their greater attunement to a range of possible causal factors.

On the other hand, it seems fairly clear that simple models are the most useful ones—at least in science—because they're easier to disprove and consequently to improve upon. Most of Aristotle's physical propositions have turned out to be demonstrably false. But Aristotle had testable propositions about the world while the Chinese did not: It was Westerners who established what the correct physical principles are. The Chinese may have understood the principle of action at a distance, but they had no means of proving it. When it was proved true, it was by Western scientists who did not initially believe in it and who were actually trying to establish that all motion was of the billiard ball type, with objects moving only because they come into contact with some other object.

Westerners' success in science and their tendency to make certain mistakes in causal analysis derive from the same source. Freedom to pursue individual goals prompts people to model the situation so as to achieve those goals, which in turn encourages modeling events by working backward from effects to possible causes. When there is systematic testing of the model, as in science, the model can be corrected. But Westerners' models tend to be limited too sharply to the goal object and its properties, slighting the possible role of context. When it is everyday life—all too often a buzzing confusion—that is being modeled, recognition of error is more difficult. A mistaken model will be difficult to correct. So despite their history of scientific-mindedness, Westerners are particularly susceptible to the Fundamental Attribution Error and to overestimating the predictability of human behavior.

As we shall see next, Westerners' preferred simplicity and Easterners' assumed complexity encompass more than their approaches to causality. Their preferences extend to the ways that knowledge is organized more generally.

CHAPTER 6

IS THE WORLD MADE UP
OF NOUNS OR VERBS?

Jorge Luis Borges, the Argentine writer, tells us that there is an ancient Chinese encyclopedia entitled *Celestial Emporium of Benevolent Knowledge* in which the following classification of animals appears: "(a) those that belong to the emperor, (b) embalmed ones, (c) those that are trained, (d) suckling pigs, (e) mermaids, (f) fabulous ones, (g) stray dogs, (h) those that are included in this classification, (i) those that tremble as if they were mad, (k) those drawn with a very fine camel's hair brush, (l) others, (m) those that have just broken a flower vase, (n) those that resemble flies at a distance.

Though Borges may have invented this classification for his own purposes, it is certainly the case that the ancient Chinese did not categorize the world in the same sorts of ways that the ancient Greeks did. For the Greeks, things belonged in the same category if they were describable by the same attributes. But the philosopher Donald Munro points out that, for the Chinese, shared attributes did not establish shared class membership. Instead, things were classed together because they were thought to influence one another through *resonance*. For example, in the Chinese system of the Five Processes, the categories spring, east, wood, wind, and green all influenced one another. Change in wind would affect all the others—in "a process like a multiple echo, without physical contact coming between any of them." Philosopher David Moser also notes that it was similarity between classes, not similarity among individual members of the same class, that was of interest to the ancient Chinese. They were simply not concerned about the relationship between a member of a class ("a horse") and the class as a whole ("horses").

In fact, for the Chinese there seems to have been a positive antipathy toward categorization. For the ancient Taoist philosopher Chuang Tzu, " . . . the problem of . . . how terms and attributes are to be delimited, leads one in precisely the wrong direction. Classifying or limiting knowledge fractures the greater knowledge." In the *Tao Te Ching* we find the following dim view of the effects of relying on categories.

> *The five colors cause one's eyes to be blind.*
> *The five tones cause one's ears to be deaf.*
> *The five flavors cause one's palate to be spoiled.*

The lack of interest in classes of objects sharing the same properties is consistent with the basic scheme that the ancient Chinese had for the world. For them, the world consisted of continuous substances. So it was a *part-whole* dichotomy that made sense to them. Finding the features shared by objects and placing objects in a class on that basis would not have seemed a very useful activity, if only because the objects themselves were not the unit of analysis. Since the Greek world was composed of objects, an *individual-class* relation was natural to them. The Greek belief in the importance of that relation was central to their faith in the possibility of accurate inductive inferences: Learning that one object belonging to a category has a particular property means that one can assume that other objects belonging to the category also have the property. If one mammal has a liver, it's a good bet that all mammals do. A focus on the *one-many*, individual-class organization of knowledge encourages induction from the single case; a part-whole representation does not.

CATEGORIES VS. RELATIONSHIPS IN MODERN THOUGHT

Once again, we have a case of very different intellectual traditions in ancient Greece and ancient China, and once again we can ask whether the mental habits of ancient philosophers resemble the perception and reasoning of ordinary people today. We might expect, based on the historical evidence for cognitive differences and our theory about the social origins of them, that contemporary Westerners would (a) have a greater tendency to categorize objects than would Easterners; (b) find it easier to learn new categories by applying rules about properties to particular cases; and (c) make more inductive use of categories, that is, generalize from particular instances of a category to other instances or to the category as a whole. We might also expect that Easterners, given their convictions about the potential relevance of every fact to every other fact, would organize the world more in terms of perceived relationships and similarities than would Westerners.

Take a look at the three objects pictured in the illustration on page 141. If you were to place two objects together, which would they be? Why do those seem to be the ones that belong together?

If you're a Westerner, odds are you think the chicken and the cow belong together. Developmental psychologist Liang-hwang Chiu showed triplets like that in the illustration to American and Chinese children. Chiu found that the American children preferred to group objects because they belonged to the "taxonomic" category, that is, the same classification term could be applied to both ("adults," "tools"). Chinese children preferred to group objects on the basis of relationships. They would be more likely to say the cow and the grass in the illustration go together because "the cow eats the grass."

Li-jun Ji, Zhiyong Zhang, and I obtained similar results comparing college students from the U.S with students from mainland China and Taiwan, using words instead of pictures. We presented participants with sets of three words (e.g., panda, monkey, banana) and asked them to indicate which two of the three were most closely related. The American participants showed a marked pref-

A   B

What goes with this? A or B

Example of item measuring preference for grouping
by categories vs. relationships.

erence for grouping on the basis of common category
membership: Panda and monkey fit into the animal category. The Chinese participants showed a preference for
grouping on the basis of thematic relationships (e.g., monkey and banana) and justified their answers in terms of
relationships: Monkeys eat bananas.

If the natural way of organizing the world for Westerners is to do so in terms of categories and the rules that
define them, then we might expect that Westerners' per-

ceptions of similarity between objects would be heavily influenced by the degree to which the objects can be categorized by applying a set of rules. But if categories are less salient to East Asians, then we might expect that their perceptions of similarity would be based more on the family resemblance among objects.

To test this possibility, Ara Norenzayan, Edward E. Smith, Beom Jun Kim, and I gave schematic figures like those shown in the illustration on page 143 to Korean, Euopean American, and Asian American participants. Each display consisted of an object at the bottom and two groups of objects above it. The participants' job was just to say which group of objects the target object seemed more similar to. You might want to make your own judgment about the objects in the illustration before reading on.

Most of the Koreans thought the target object was more similar to the group on the left, whereas most of the European Americans thought the object was more similar to the group on the right. The target object bears a more obvious family resemblance to the group on the left, so it's easy to see why the Koreans would have thought the object was more similar to that group, and on average they did so 60 percent of the time. But there is a simple, invariant rule that allows you to place the target object into a category that it shares with the group on the right. The rule is "has a straight (as opposed to curved) stem." European Americans typically discovered such rules and, 67 percent of the time, found the target object to be more similar to the group with which it shared the rule-based category. Asian American judgments were in between but more similar to those of the Koreans.

Group 1   Group 2

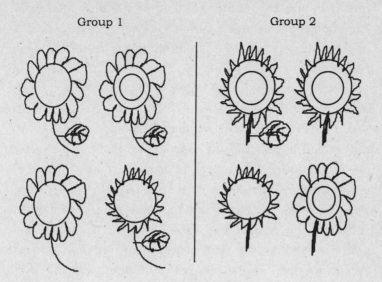

Target Object

Example of item measuring whether judgments of similarity
are based on family resemblance or rules.

∞

Categories are sometimes learned by applying rules to features. We come to know that rabbits are mammals because
we are taught a rule that animals that nurse their young
are mammals. (That's true for categories defined formally,
in any case. Actually, most people probably learn what

mammals are by ostention: "that rabbit is a mammal," "that lion is a mammal." The "folk" category that is learned is then induced from the common properties observed—fur-bearing, four-footed, etc.)

Explicit modeling or rule-making seems to be less characteristic of the causal explanations of East Asians than of Westerners. If Asians are less likely to use rules to understand the world, and less likely to make use of categories, they might find it particularly hard to learn categories by applying explicit rules to objects. In order to test this possibility, Ara Norenzayan and his colleagues showed color cartoon figures like those rendered in black and white in the illustration on page 145 to East Asian, Asian American, and European American students at the University of Michigan. We told participants that they would be learning how to classify the animals as being either from Venus or from Saturn.

We told participants that an animal was from Venus if it had any three of five features: curly tail, hooves, long neck, mouth, and antennae ears. Otherwise, the creature was from Saturn. The animal on the left at the top (seen as blue by participants) meets the criteria for being from Venus; the one on the right (seen as red) doesn't and has to be put in the Saturn category. After participants had learned how to classify animals correctly, we tested how much control they had over the categories by showing them new animals and seeing how fast and accurately they could classify them. The new animals included two types that resembled previously seen ones. Some animals were "positive matches": They looked like an animal participants had seen before during the training trials and they

TRAINING PHASE

Known Venus   Known Saturn

TEST PHASE: Positive Match   TEST PHASE: Negative Match

Venus   Saturn

Example of cartoon animals used for study of
ease of learning categories based on rules.

belonged to the same category in terms of the rules concerning their features. Other animals were "negative
matches": They looked like an animal that had been seen
before, but in terms of the rules, they belonged to a different category from the one seen in training. The animal on
the lower left is a positive match for the one on the left
above: It looks like the one categorized as being from

Venus and the rules also indicate that it is. The one on the lower right is a negative match: It looks like the Venus animal but the rules say it's not.

The Asian participants took longer to make their judgments about whether the animal was from Venus or Saturn than either the European Americans or Asian Americans. The three groups of participants were equally fast and equally accurate for the positive matches, for which both memory for the previously seen example and correct applications of the rules defining the category would produce the correct answer. But for the negative matches, which could be classified correctly only if the rules were remembered and applied correctly, Asian participants made twice as many classification errors as either European Americans or Asian Americans did. Categorization by rules seems not to come as easily to Easterners as to Westerners.

Which of the two arguments below, both ending in the conclusion "rabbits have enzyme Q in their blood," seems more convincing to you? Why?

(1)	(2)
Lions have enzyme Q in their blood	Lions have enzyme Q in their blood
Tigers have enzyme Q in their blood	Giraffes have enzyme Q in their blood
---	---
Rabbits have enzyme Q in their blood	Rabbits have enzyme Q in their blood

Most Westerners who have been asked this sort of question say that argument 2 is better. They give as their reason some version of a "diversity" or "coverage" argument. Lions and tigers are rather similar animals in many ways, so they don't cover the mammal category, to which rabbits belong, very well. Lions and giraffes give better coverage of the mammal category because they're more different from each other. Now consider the arguments below, both ending in the conclusion "mammals have enzyme Q in their blood." Which seems more convincing to you?

(1)	(2)
Lions have enzyme Q in their blood	Lions have enzyme Q in their blood
Tigers have enzyme Q in their blood	Giraffes have enzyme Q in their blood
Mammals have enzyme Q in their blood	Mammals have enzyme Q in their blood

Again, most Westerners say the second argument is more convincing and give as their reason that the coverage of the mammal category is better for the second argument than for the first.

Incheol Choi, Edward E. Smith, and I gave problems like those above to Korean and American college students. Koreans, but not Americans, were more likely to prefer the second argument when the category was mentioned in the conclusion. For Koreans, the mammal category was not salient unless it was highlighted by actually referring to it. As a result, the diversity principle was more important to their inferences when they were explicitly reminded that the objects in question were mammals. One likely consequence of the low salience of categories for Easterners is that they do not fuel inductive inferences for Easterners as much as for Westerners.

GROWING UP IN A WORLD OF
OBJECTS VS. RELATIONSHIPS.

How is it possible that Easterners today have relatively little interest in categories, find it hard to learn new categories by applying rules about properties, and make little
spontaneous use of them for purposes of induction? Why
are they so much more inclined to consider relationships
in their organization of objects than Westerners are?
Surely not just because ancient Chinese philosophers had
little use for categories and were more interested in part-
whole relationships and thematic resemblances than in
category-member classifications. It seems dubious that
philosophers' concerns would have affected judgments
about everyday objects even by their contemporaries. If
relationships, and not categories, are relatively important
to East Asians today, there must be factors that still operate in the socialization of children that prompt such different styles of perception and reasoning. Before looking
for such factors, let's consider some important differences
between categories and relationships.

Categories are denoted by nouns. It seems obvious
that nouns would be easier for a young child to learn than
verbs. All you have to do to learn that the animal you just
saw is a "bear" is to notice its distinctive features—huge
size, large teeth and claws, long fur, ferocious appearance—and you can store that object away with its label.
The label is then available for application to any other
object having that set of properties.

Relationships, on the other hand, involve, tacitly or
explicitly, a verb. Learning the meaning of a transitive verb

normally involves noticing two objects and some kind of action that connects them in some way. "To throw" means to use your arm and hand to move an object through the air to a new location. Merely pointing at the action does not guarantee that someone will know what you're referring to.

Because of their relative ambiguity, it's harder to remember verbs; verbs are more likely to be altered in meaning than nouns when a speaker communicates to another person or when one person paraphrases what another has said; and it's harder to correctly identify verbs than nouns when they're translated from one language to another. Moreover, the meaning of verbs, and other terms that describe relations, differs more across different languages than simple nouns do. "Verbs," says cognitive psychologist Dedre Gentner, "are highly reactive; nouns tend to be inert."

Given these differences between nouns and verbs, it is scarcely surprising that Gentner finds that children learn nouns much more rapidly than they learn verbs. In fact, toddlers can learn nouns at rates of up to two per day. This is much faster than the rate at which they learn verbs.

Gentner quite reasonably guessed that the large noun advantage would be universal. But it turns out not to be. Developmental psycholinguist Twila Tardif and others have found that East Asian children learn verbs at about the same rate as nouns and, by some definitions of what counts as a noun, at a significantly faster rate than nouns. There are several factors that might underlie this dramatic difference.

First, verbs are more salient in East Asian languages

than in English and many other European languages. Verbs in Chinese, Japanese, and Korean tend to come either at the beginning or the end of sentences and both are relatively salient locations. In English, verbs are more commonly buried in the middle.

Second, recall from chapter 3 the father I overheard quizzing his child about the properties of pants. Western parents are noun-obsessed, pointing objects out to their children, naming them, and telling them about their attributes. Strange as it may seem to Westerners, Asians don't seem to regard object naming as part of the job description for a parent. Developmental psychologists Anne Fernald and Hiromi Morikawa went into the homes of Japanese and Americans having infants either six, twelve, or nineteen months old. They asked the mothers to clear away the toys from a play area and then they introduced several that they had brought with them—a stuffed dog and pig and a car and a truck. They asked the mothers to play with the toys with their babies as they normally would. They found big differences in the behavior of mothers even with their youngest children. American mothers used twice as many object labels as Japanese mothers ("piggie," "doggie") and Japanese mothers engaged in twice as many social routines of teaching politeness norms (empathy and greetings, for example). An American mother's patter might go like this: "That's a car. See the car? You like it? It's got nice wheels." A Japanese mother might say: "Here! It's a vroom vroom. I give it to you. Now give this to me. Yes! Thank you." American children are learning that the world is mostly a place with objects, Japanese children that the world is mostly about relationships.

Third, we know that naming objects that share a common set of properties results in infants' learning a category formed of objects sharing those features. Naming objects sharing features also prompts them to attend to features that would allow them to form other categories based on similar sets of properties. Developmental psychologists Linda Smith and her colleagues randomly assigned seventeen-month-old children either to a control condition or to a condition in which, for nine weeks, they repeatedly played with and heard names for members of unfamiliar object categories that were defined by shape: for example, "cup." This taught the toddlers to attend to shape and to form categories for objects—even those seen outside the experimental setting—that could be grouped on the basis of some set of defining features. The result was that trained children showed a dramatic increase in acquisition of new object names during the course of the study.

Fourth, generic nouns (that is, category names) in English and other European languages are often marked by syntax. When the conversation turns to waterfowl, you can say "a duck," "the duck," "the ducks," or "ducks." The last term is a generic one and the syntax tells you this. It's normally obligatory to indicate whether you're speaking about an object or a class of objects, though sometimes the context can do the job. But in Chinese and other Sinitic languages, contextual and pragmatic cues can be the only kinds of cues the hearer has to go on. The presence of a duck that has just waddled over from a pond to beg food, for example, would indicate that it is "the duck" one is talking about, rather than "a duck," "the ducks," or "ducks." Developmental psychologists Susan Gelman and

Twila Tardif studied English-speaking mothers and Mandarin Chinese–speaking mothers and found that, across a number of contexts, generic utterances were more common for the English-speaking mothers.

Finally, there is direct evidence that Eastern children learn how to categorize objects at a later point than Western children. Developmental psycholinguists Alison Gopnik and Soonja Choi studied Korean-, French-, and English-speaking children beginning when they were one and a half years old. They found that object-naming and categorization skills develop later in Korean speakers than in English and French speakers. The investigators studied means-ends judgments (for example, figuring out how to take things out of a container), and categorization, which they studied by showing children four objects of one kind and four of another, such as four flat, yellow rectangles and four small human figures, and telling them to "fix these things up," that is, put them together in some way that makes sense. English- and French-speaking toddlers mastered the means-end tasks and the categorization tasks at about the same age. Korean toddlers learned categorization almost three months later than means-end abilities.

DISPOSITIONS, STABILITY, AND CATEGORIES

The ancient Greeks were fond of categories and used them as the basis for discovery and application of rules. They also believed in stability and understood both the physical and social worlds in terms of fixed attributes or dispositions. These are not unrelated facts, nor is it a coincidence that the ancient Chinese were uninterested in categories, believed in change, and understood the behavior of both physical and social objects as being due to the interaction of the object with a surrounding field of forces.

If the world is a stable place, then it is worthwhile trying to develop rules to understand it and refining the categories to which the rules apply. Many of the categories used to understand the world refer to presumed qualities of the object: hardness, whiteness, kindness, timidity. Easterners of course use such categories as well, but they are less likely to abstract them away from particular objects: There is the whiteness of the horse or the whiteness of the snow in ancient Chinese philosophy, but not whiteness as an abstract, detachable concept that can be applied to almost anything. In the Western tradition, objects have essences composed of mix-and-match abstract qualities. These essences allow for confident predictions about behavior independent of context. In the Eastern tradition, objects have concrete properties that interact with environmental circumstances to produce behavior. There was never any interest in discussing abstract properties as if they had a reality other than being a characteristic of a particular object.

Most importantly, the dispositions of objects are not necessarily stable for Easterners. In the West, a child who performs poorly in mathematics is likely to be regarded as having little math ability or perhaps even as being "learning disabled." In the East, such a child is viewed as needing to work harder, or perhaps her teacher should work harder, or maybe the setting for learning should be changed.

∽

The obsession with categories of the either/or sort runs through Western intellectual history. Dichotomies abound in every century and form the basis for often fruitless debates: for example, "mind-body" controversies in which partisans take sides as to whether a given behavior is best understood as being produced by the mind independent of any biological embodiment, or as a purely physical reaction unmediated by mental processes. The "nature-nurture" controversy is another debate that has often proved to generate more heat than light. As evolutionary biologist Richard Alexander has pointed out, nearly all behaviors that are characteristic of higher order mammals are determined by both nature and nurture. The dichotomy "emotion-reason" has obscured more than it has revealed. As Hume said, "reason is and ought to be the slave of passion"; it makes sense to separate the two for purposes of analysis only. And it's been suggested that the distinction between "human" and "animal" insisted upon by Westerners made it particularly hard to accept the concept of evolution. In most Eastern systems, the soul can take the form of any animal or even God. Evolution was never controversial in the East because there was never an assumption that humans sat atop a chain of being and somehow had lost their animality.

Throughout Western intellectual history, there has been a conviction that it is possible to find the necessary and sufficient conditions for any category. A square is a two-dimensional object with four sides of equal length and four right angles. Nothing lacking these properties can be a square and anything having those properties is definitely a square. Ludwig Wittgenstein, in his *Philosophical*

Investigations, brought the whole necessity-and-sufficiency enterprise crashing to earth in the West. Wittgenstein argued to the satisfaction (or rather, dismay) of even the most analytic of Western philosophers that establishing necessary and sufficient conditions for any complex or interesting category, such as a "game" or a "government" or an "illness," was never going to be possible. A thing can be a game even if it is not fun, even if played alone, even if its chief goal is to make money. A thing is not necessarily a game even if it is fun or is a nonproductive activity engaging several people in pleasurable interaction. Wittgenstein's sermon would never have been needed in the East. The pronouncement that complex categories cannot always be defined by necessary and sufficient conditions would scarcely have been met with surprise.

Is It Language That Does the Job?

Given the substantial differences in language usage between Easterners and Westerners, is it possible that it is merely language that is driving the differences in tendency to organize the world in terms of verbs vs. nouns? Are the findings about knowledge organization simply due to the fact that Western languages encourage the use of nouns, which results in categorization of objects, and Eastern languages encourage the use of verbs, with the consequence that it is relationships that are emphasized? More generally, how many of the cognitive differences documented in this book are produced by language?

There are in fact a remarkable number of parallels

between the sorts of cognitive differences discussed in this book and differences between Indo-European languages and East Asian languages. The parallels are particularly striking because East Asian languages, notably Chinese and Japanese, are themselves so different in many respects, yet nevertheless share many qualities with one another that differentiate them from Indo-European languages.

In addition to the practices already discussed—pointing and naming, location of verbs in sentences, marking of nouns as generic, and so on—there are several ways in which language usage maps onto differences in category usage.

The Western concern with categories is reflected in language. "Generic" noun phrases are more common for English speakers than for Chinese speakers, perhaps because Western languages mark in a more explicit way whether a generic interpretation of an utterance is the correct one. In fact, in Chinese there is no way to tell the difference between the sentence "squirrels eat nuts" and "this squirrel is eating the nut." Only context can provide this information. English speakers know from linguistic markers whether it is a category or an individual that is being talked about.

Greek and other Indo-European languages encourage making properties of objects into real objects in their own right—simply adding the suffix "ness" or its equivalent. The philosopher David Moser has noted that this practice may foster thinking about properties as abstract entities that can then function as theoretical explanations. Plato actually thought that these abstractions had a greater reality than the properties of objects in the physical world. This degree of theorizing about abstractions was never characteristic of Chinese philosophy.

East Asian languages are highly "contextual." Words (or phonemes) typically have multiple meanings, so to be understood they require the context of sentences. English words are relatively distinctive and English speakers in addition are concerned to make sure that words and utterances require as little context as possible. The linguistic anthropologist Shirley Brice Heath has shown that middle-class American parents quite deliberately attempt to decontextualize language as much as possible for their children. They try to make words understandable independent of verbal context and to make utterances understandable independent of situational context. When reading to a child about a dog, the parent might ask the child what the animal is ("A doggie, that's right") and who has a dog ("Yes, Heather has a dog"). The word is detached from its naturally occurring context and linked to other contexts where the word has a similar meaning.

Western languages force a preoccupation with focal objects as opposed to context. English is a "subject-prominent" language. There must be a subject even in the sentence "It is raining." Japanese, Chinese, and Korean, in contrast, are "topic-prominent" languages. Sentences have a position, typically the first position, that should be filled by the current topic: "This place, skiing is good." This fact places an alternative interpretation on our finding that, after viewing underwater scenes, Americans start with describing an object ("There was a big fish, maybe a trout, moving off to the left") whereas Japanese start by establishing the context ("It looked like a pond"). While not obligatory from a grammatical standpoint, an idiomatic Japanese sentence starts with context and topic rather than jumping immediately to a subject as is frequently the case in English.

For Westerners, it is the self who does the acting; for Easterners, action is something that is undertaken in concert with others or that is the consequence of the self operating in a field of forces. Languages capture this different sort of agency. Recall that there are many different words for "I" in Japanese and (formerly, at any rate) in Chinese, reflecting the relationship between self and other. So there is "I" in relation to my colleague, "I" in relation to my spouse, etc. It is difficult for Japanese to think of properties that apply to "me." It is much easier for them to think of properties that apply to themselves in certain settings and in relation to particular people. Grammar also reflects a different sense of how action comes about. Most Western languages are "agentic" in the sense that the language conveys that the self has operated on the world: "He dropped it." (An exception is Spanish.) Eastern languages are in general relatively nonagentic: "It fell from him," or just "fell."

A difference in language practice that startles both Chinese speakers and English speakers when they hear how the other group handles it concerns the proper way to ask someone whether they would like more tea to drink. In Chinese one asks "Drink more?" In English, one asks "More tea?" To Chinese speakers, it's perfectly obvious that it's tea that one is talking about drinking more of, so to mention tea would be redundant. To English speakers, it's perfectly obvious that one is talking about drinking the tea, as opposed to any other activity that might be carried out with it, so it would be rather bizarre for the question to refer to drinking.

According to linguistic anthropologists Edward Sapir and Benjamin Whorf, the differences in linguistic structure between languages are reflected in people's habitual thinking processes. This hypothesis has moved in and out of favor among linguists and psychologists over the decades, but it is currently undergoing one of its periods of greater acceptance. Some of our evidence about language and reasoning speaks directly to the Sapir-Whorf hypothesis.

Recall that Li-jun Ji, Zhiyong Zhang, and I examined whether language per se affects the way people categorize objects. We gave word triplets (for example, panda, monkey, banana) to Chinese and American college students and asked them to indicate which two of the three were most closely related. The Chinese students were either living in the U.S. or in China and they were tested either in English or in Chinese.

If the Sapir-Whorf hypothesis is correct, then it ought to make a difference which language the bilingual Chinese are tested in. They should be more likely to prefer relationships (monkey, banana) as the basis for grouping when tested in Chinese and more likely to prefer taxonomic category (panda, monkey) when tested in English. But there are different ways of being bilingual. Psycholinguists make a distinction between what they call "coordinate" bilinguals and "compound" bilinguals. Coordinate bilinguals are people who learn a second language relatively late in life and for whom its use is confined to a limited number of contexts. Mental representations of the world supposedly can be different in one language than in the other for such people. Compound bilinguals are people for whom the second language is learned early and is used in many contexts. Mental representations for such people should be fused, since the languages are not used for different functions or used exclusively in different settings. We tested both types of bilinguals. People from China and Taiwan could be expected to be coordinate bilinguals because they typically learn English relatively late and its use is confined mostly to formal school contexts. People from Hong Kong and Singapore would be more likely to be compound bilinguals because they learn English relatively early and use it in more contexts. In addition, these societies, especially Hong Kong, are highly Westernized.

If language makes a difference to understanding of the world because different languages underlie different mental representations, we would expect to find the Sapir-Whorf hypothesis supported: The coordinate bilinguals, at least, should group words differently when tested in Chinese than when tested in English. If language makes a difference because structural features of the language compel different thinking processes, then we might expect even the compound bilinguals to group words differently when tested in Chinese than when tested in English. And, of course, if language is not important to cognitive tasks such as our grouping one, then we would expect no effect of language for either group.

The results could not have been more unequivocal. First, there were marked differences between European Americans tested in English and coordinate Chinese speakers tested in Chinese, whether in China or in the U.S. Americans were twice as likely to group on the basis of taxonomic category as on the basis of relationships. Mainland and Taiwanese Chinese tested in their native language were twice as likely to group on the basis of relationships as on the basis of taxonomic category and this was true whether they were tested in their home countries or in the U.S. Second, the language of testing did make a big difference for the mainland and Taiwanese Chinese. When tested in English, they were much less likely to group on the basis of relationships. It thus appears that English subserves a different way of representing the world than Chinese for these participants.

But matters were quite different for compound bilinguals from Hong Kong and Singapore. First, their groupings were shifted in a substantially Western direction: They were still based on relationships more than on taxonomic category, but the preference was much weaker for them than for the coordinate Chinese and Taiwanese speakers. More importantly, it made precisely no difference for the compound speakers whether they were tested in Chinese or in English.

The results are clear in their implications. There is an effect of culture on thought independent of language. We know this because both the coordinate Chinese speakers and the compound Chinese speakers group words differently from Americans regardless of language of testing. The differences between coordinate and compound speakers also indicate a culture difference independent of language. The compound speakers from Westernized regions are shifted in a Western direction—and to the same extent regardless of language of testing. There is also clearly an effect of language independent of culture—but only for the coordinate speakers from China and Taiwan. They respond very differently depending on whether they are tested in Chinese or in English.

A tentative answer to the Sapir-Whorf question as it relates to our work—and it must be very tentative because we have just been discussing a couple of studies dealing with a single kind of mental process—is that language does indeed influence thought so long as different languages are plausibly associated with different systems of representation.

So there is good evidence that for East Asians the world is seen much more in terms of relationships than it is for Westerners, who are more inclined to see the world in terms of static objects that can be grouped into categories. Child-rearing practices undoubtedly play a role in producing these very different visions. East Asian children have their attention directed toward relationships and Western children toward objects and the categories to which they belong. Language probably plays a role, at least in helping to focus attention, but probably also in stabilizing the different orientations throughout life. There appears to be nothing about the structure of language, though, that actually forces description in terms of categories versus relationships.

As we will see next, the very different approaches to

understanding the world don't stop with the organization of knowledge. The decontextualization and object emphasis favored by Westerners, and the integration and focus on relationships by Easterners, result in very different ways of making inferences.

CHAPTER 7

"CE N'EST PAS LOGIQUE" OR "YOU'VE GOT A POINT THERE"?

...The most striking difference between the
traditions at the two ends of the civilized world
is in the destiny of logic. For the West, logic has
been central and the thread of transmission has
never snapped ...
—PHILOSOPHER ANGUS GRAHAM

...It is precisely because the Chinese mind is
so rational that it refuses to become rationalistic
and ... to separate form from content.
—PHILOSOPHER SHU-HSIEN LIU

The aim of the Chinese classical education has
always been the cultivation of the reasonable

man as the model of culture. An educated man
should, above all, be a reasonable being, who is
always characterized by his common sense, his
love of moderation and restraint, and his hatred
of abstract theories and logical extremes.
— LITERARY CRITIC LIN YUTANG

To argue with logical consistency . . . may not
only be resented but also be regarded as immature.
— ANTHROPOLOGIST NOBUHIRO NAGASHIMA

Hard as it is for the Westerner to understand, there were
only two short-lived movements of little influence in the
East that shared the spirit of logical inquiry that has
always been common in the West. These were the *Ming jia*
(Logicians) and the Mohists, or followers of Mo-tzu, both
of the classical period in antiquity. The Logicians in fact
made little progress toward a formal logic, though, unlike
adherents of all other traditions of Chinese philosophy,
they were interested in knowledge for its own sake. The
Mo-tzu tradition embraced several logical concerns, chief
among them the ideas of necessary and sufficient conditions and the principle of noncontradiction and the law of
the excluded middle. Nevertheless, even the Mohists
stopped short of producing a rigorous system of logical
inference. Moreover, despite the Mohists' advances in
geometry, they never formalized it in Western fashion and
never developed a set of foundational principles that
would allow logical derivation of solutions.

The best explanation for the Greeks' concern with

logic is that they saw its utility in argumentation. So it seems to be no coincidence that Mo-tzu both was concerned with logic and believed that argumentation was valuable for clarifying propositions and for helping to distinguish between right and wrong. Mo-tzu wanted to develop ways of maximizing the common good and he actually developed a rough version of cost-benefit analysis. These facts put him closer in spirit to modern Western philosophy than to either ancient Chinese or ancient Greek philosophy. Even in these aspects of his work, however, he retained an Eastern orientation. Like other Chinese philosophers he made no distinction between the truth of a proposition and its morality, a position that, whatever its effects on ethics, is deadly for logic.

By the first millennium A.D. there were essentially no traces of a logical approach to understanding the world. Instead there was a trust in sense impressions and common sense. And there was never, even among the Logicians and Mohists, a willingness to accept arguments that flew in the face of experience—unlike the Greeks, who sometimes seemed quite delighted to deny the evidence of the senses. As we will see, the Chinese remain far more committed to reasonableness than to reason.

LOGIC VS. EXPERIENCE

Integrally related to the lack of interest in logic in the East has been a distrust of *decontextualization*, that is, of considering the structure of an argument apart from its content, as well as a distaste for making inferences on the basis of underlying abstract propositions alone. Two studies by Ara Norenzayan, Edward E. Smith, Beom Jun Kim, and me show how this remains true for ordinary people in twenty-first-century Asia.

Consider the following two deductive arguments. Is one more convincing than the other?

1. All birds have ulnar arteries.
   Therefore all eagles have ulnar arteries.
2. All birds have ulnar arteries.
   Therefore all penguins have ulnar arteries.

(No need to know what an ulnar artery is. It's in effect a "blank" property, used so that real-world knowledge can't intrude into the evaluation of a deductive argument.)

One way to measure the extent to which people spontaneously rely on formal logic versus experiential knowledge in reasoning is to examine how they "project" properties—"ulnar arteries" in the above example—from *superordinate* categories (birds) to *subordinate* categories (eagles, penguins). Notice that the two arguments have identical premises but their conclusions vary in how typical the target bird is. Eagles are more typical birds than penguins. If you are in pure logical mode when you evaluate propositions like those above, you will supply the implicit middle premises of the arguments ("All eagles are birds," and "All penguins are birds"). People who do this would find the two arguments equally convincing. But people often find arguments to a typical instance to be more convincing than arguments to atypical ones. Prior experience makes them more comfortable with regarding eagles as birds than regarding penguins as birds.

We asked Korean, Asian American, and European American participants to evaluate the convincingness of twenty such arguments, ten with typical targets like eagles in the conclusion and ten with atypical targets like penguins. We found that Koreans were more convinced by typical arguments than by atypical arguments. European Americans, in contrast, were almost equally convinced by typical and atypical arguments. Asian Americans' responses were in between those of European Americans and Koreans.

Consider the arguments below. Which ones seem to you to be logically valid?

> Premise 1: No police dogs are old.
> Premise 2: Some highly trained dogs are old.
> Conclusion: Some highly trained dogs are not police dogs.

> Premise 1: All things that are made from plants are good for health.
> Premise 2: Cigarettes are things that are made from plants.
> Conclusion: Cigarettes are good for health.

> Premise 1: No A are B.
> Premise 2: Some C are B.
> Conclusion: Some C are not A.

The first argument is meaningful and has a plausible conclusion, the second argument is meaningful but its conclusion is not plausible, and the third argument is so abstract that it has no real meaning at all. But all three arguments are logically valid.

People are more likely to be correct in their judgments about the logical validity of arguments when the argument is meaningful and its conclusion is plausible. They are least likely to be correct when the argument is meaningful and its conclusion is implausible. We presented Korean and American college students with arguments that were either valid or invalid and that had conclusions that were either plausible or implausible and asked them to evaluate whether or not the conclusion followed logically from the premises for each argument. We examined four different types of syllogisms, ranging from the very simple *modus ponens* (If A is the case, then B is the case; A is the case; therefore B is the case) to the difficult structure in the third example above.

Both Koreans and Americans were more likely to rate syllogisms with plausible conclusions as valid. As expected, though, Koreans were more influenced by plausibility than Americans. There is no question of this difference being due to the Korean participants being less capable of performing logical operations than the American participants. Koreans and Americans made an equal number of errors on the purely abstract syllogisms. The difference between the two groups would seem to be that Americans are simply more in the habit of applying logical rules to ordinary events than Koreans and are therefore more capable of ignoring the plausibility of the conclusions.

꙳

East Asians, then, are more likely to set logic aside in favor of typicality and plausibility of conclusions. They are also more likely to set logic aside in favor of the desirability of conclusions.

William McGuire showed that when people are asked to judge the probability of events that bear a logical relation to one another, their probability judgments move into line with one another in such a way as to increase the logical coherence of the beliefs taken as a whole. For example, McGuire asked people how likely they thought it was that (a) there would be a drought that summer; (b) drought would mean that the beaches would be polluted because of undiluted runoff; (c) if the beaches were polluted, the authorities would close them; and (d) the beaches would be closed. McGuire found that, over time, the logical consistency among people's beliefs about the related propositions increased, merely by virtue of asking them to think about the likelihood that they were true. Two weeks after making their probability estimates for a number of items like those above, the probabilities participants gave for the various propositions were more in line with logical requirements than they had been initially, before they had had time to think about it. So although people didn't want the beaches to be closed, after thinking about it for a while in relation to other propositions that had some significant likelihood of being true, and which implied directly or indirectly that the beaches might be closed, they became more pessimistic about their summer seaside plans.

Ara Norenzayan and Beom Jun Kim guessed that East

Asians would be less likely to have their beliefs moved in an unpleasant direction by pondering information that implied some undesired outcome—because East Asians are not so accustomed to applying logic to everyday life events and therefore might be able to cling successfully to beliefs that were countered by other propositions they were asked to think about. They gave Korean and American students propositions that had a logical relation to one another, but mixed the propositions in with many others so that it was unlikely that participants would realize that consistency among their probability judgments was being tested. Spread out through the questionnaire, for example, were the following propositions:

The price of dining out will increase.
If stricter health codes for restaurants will increase the cost of hiring new staff, then the price of dining out will increase.
Stricter health codes for restaurants will increase the cost of hiring new staff.

Some of the propositions were positive: for example, "more poor people will be able to get enough food to stay healthy." Other propositions, like the one above about the increased price of dining out, were unattractive. Norenzayan and Kim asked participants about the probabilities they assigned to the various propositions at two different times: immediately after they read each proposition and a few minutes after they had read all the propositions.

Korean and American participants' beliefs showed equal consistency the first time they were tested and the consistency of the two groups was equal—and greater for both groups—the second time around for the positive propositions. But the Americans also moved more in a consistent direction for the negative propositions and the Koreans did not. Apparently when logical push came to desirable shove, the logical implications of some beliefs for others were less likely to affect the probability judgments of Koreans than those of Americans.

Either/Or vs. Both/And

Which of the following two groups of proverbs most appeals to you—the first three or the second three?

> Half a loaf is better than none.
> One against all is certain to fall.
> "For example" is no proof.

> Too humble is half-proud.
> Beware of your friends, not your enemies.
> A man is stronger than iron and weaker than a fly.

The second set of proverbs express apparent contradictions: Humble is not proud and friends are just the sort of person you shouldn't have to be wary of. The first set may or may not seem very pithy, but none embody contradictions. Kaiping Peng and I found that the second type of proverbs were more common in a Chinese compendium of proverbs than in an American collection. When we asked students at the University of Michigan and at Beijing University to rate how much they liked the proverbs, we found that the Chinese students had a preference for the proverbs with contradictions and the Americans had a preference for the proverbs without them. To make sure that it wasn't familiarity with the proverbs that was producing the differences, we conducted a study using Yiddish proverbs. We obtained similar results: Americans and Chinese were equally fond of the proverbs without contradictions, but the Chinese liked the ones containing contradictions more than did the Americans. (Here again we found a similarity between Far East and Near East traditions: Yiddish proverbs were fully as likely to embody contradictions as Chinese.)

The reasons for these differences in preference for contradiction are deep. There is a style of reasoning in Eastern thought, traceable to the ancient Chinese, which has been called *dialectical*, meaning that it focuses on contradictions and how to resolve them or transcend them or find the truth in both. At the risk of doing violence to the spirit of dialecticism, which does not make use of hard and fast rules about reasoning, we can describe three principles that are important to it, which Kaiping Peng has articulated.

The Principle of Change The Eastern tradition of thought emphasizes the constantly changing nature of reality. The world is not static but dynamic and changeable. Being in a given state is just a sign that the state is about to change. Because reality is in constant flux, the concepts that reflect reality are fluid and subjective rather than being fixed and objective.

The Principle of Contradiction Because the world is constantly changing, oppositions, paradoxes, and anomalies are continuously being created. Old and new, good and bad, strong and weak exist in everything. In fact opposites complete each other and make each other up. Taoists see the two sides of any apparent contradiction existing in an active harmony, opposed but connected and mutually controlling. "Tao is conceived as both 'is' and 'is not.' " As the founder of the Taoist School, Lao-tzu, put it: "When the people of the world all know beauty as beauty, there arises the recognition of ugliness; when they all know the good as good, there arises the recognition of evil. And so, being and nonbeing produce each other . . ." Or as Mao Tse-tung, longtime Chinese dictator who regarded himself as a philosopher and poet as well as a politician and soldier, wrote: " . . . On the one hand [opposites] are opposed to each other, and on the other they are interconnected, interpenetrating, interpermeating and interdependent, and this character is described as identity."

The Principle of Relationship, or Holism As a result of change and opposition, nothing exists in an isolated and independent way, but is connected to a multitude of different things. To really know a thing, we have to know all its relations, like individual musical notes embedded in a melody.

The three principles of dialectical reasoning are related. Change produces contradiction and contradiction causes change; constant change and contradiction imply that it is meaningless to discuss the individual part without considering its relationships with other parts and prior states. The principles also imply another important tenet of Eastern thought, which is the insistence on finding the Middle Way between extreme propositions. There is a strong presumption that contradictions are merely apparent and to believe that "A is right and B is not wrong either." This stance is captured by the Zen Buddhist dictum that "the opposite of a great truth is also true."

To many Westerners, these notions may seem reasonable and even familiar. Moreover, there is a dialectical tradition of a kind that has held a place in Western thought since the time of Kant, Fichte, and Hegel. (Though the Hegelian or Marxist dialectic, with its emphasis on thesis, antithesis, and synthesis, has been held to be more "aggressive" than the Eastern variety because the effort is always toward obliterating the contradiction rather than accepting it or transcending it or using it to understand some state of affairs better.)

But Westerners tend not to be aware of the strength of their commitment to some logical principles that conflict directly with the spirit of Eastern dialecticism. These include the *law of identity*, which holds that a thing is itself and not some other thing, and the *law of noncontradiction*, which holds that a proposition can't be both true and false. The Western insistence on this pair of logical principles and the Eastern spirit of dialecticism are, on the surface at least, in direct opposition to each other.

The law of identity insists on cross-situational consistency: A is A regardless of the context. The law of noncontradiction demands that a proposition and its negation cannot both be true: A and not-A are impossible. The principle of holism, in contrast, indicates that a thing is different in one context than in another and the principle of change indicates that life is a constant passing from one state of being to another, so that to be is not to be and not to be is to be. A man is literally a different person in the family than in his role as a businessman; wealth means poverty is around the corner.

Modern East Asians are of course perfectly well aware of the same logical principles that Westerners hold dear and make use of logic in some contexts, as we have just seen. But in the East Asian view, the law of noncontradiction applies only to the realm of concepts and abstractions. The rejection of conclusions because they seem formally contradictory can be mistaken, because concepts are merely reflections of things and it can sometimes be more sensible to admit that an apparent contradiction exists than to insist that either one state of affairs or its opposite is the true one.

The differences in the two stances toward contradiction have some interesting consequences for reasoning in many domains.

Peng and I asked Chinese and American graduate students at the University of Michigan to read stories about conflicts between people and about conflicts between a person's own opposed impulses. One story reported a value conflict between mothers and daughters and another described a conflict between wanting to have fun and having to work hard in school. We asked participants to analyze these conflicts and we coded them as to whether they were Middle Way, dialectical resolutions or nondialectical resolutions. A dialectical response usually included sentences that attributed the cause of the problem to both sides and attempted to reconcile opposing views by compromise or transcendence. A response such as "both the mothers and the daughters have failed to understand each other" would have been coded as dialectical, as would an answer that pointed out that in the not-too-distant future it was likely that the two would come to see eye-to-eye. Nondialectical responses generally found exclusive fault with one side or the other.

For the mother-daughter conflict, 72 percent of Chinese answers were scored as dialectical and only 26 percent of American responses were. For the school vs. fun conflict, about half the Chinese responses were dialectical, but only about 12 percent of American responses were. In short, most of the Chinese responses tried to find a Middle Way. Most of the American responses demanded change solely in one direction.

In another study, Peng and I examined Easterners' and Westerners' preference for logical vs. dialectical arguments. We asked participants which of two arguments they preferred against Aristotle's assumption that a heavier object falls to the ground first. All of the participants were graduate students in the natural sciences at the University of Michigan, but none of them were physicists. Each argument began with: "Aristotle believed that the heavier a body is, the faster it falls to the ground. However, such an assumption might be false."

The first, logical argument, essentially Galileo's classical one, continued: "Suppose that we have two bodies, a heavy one called H and a light one called L. Under Aristotle's assumption, H will fall faster than L. Now suppose that H and L are joined together. . . . Now what happens? Well, L plus H is heavier than H so by the initial assumption it should fall faster than H alone. But in the joined body . . . L [is lighter and] will act as a 'brake' on H, and L plus H will fall slower than H alone. Hence it follows from the initial assumption that L plus H will fall both faster and slower than H alone. Since this is absurd, the initial assumption must be false."

The second, holistic or dialectical argument continued: " . . . this assumption is based on a belief that the physical object is free from any influences of other contextual factors . . . which is impossible in reality. Suppose that we have two bodies, a heavy one called H and a light one called L. If we put two of them in two different conditions, such as H in windy weather (W) and L in quiet weather (Q) . . . W or Q, would make a difference. Since these kinds of contextual influences always exist, we conclude that the initial assumption must be false."

We also asked participants which of two arguments they preferred for the existence of God, a logical one or a holistic one. The "logical" argument was a version of the ancient "cosmological" one. "Whatever exists must have a cause. . . . In moving from effects to causes, therefore, we must have two options. One is to go on tracing an infinite succession . . . without any ultimate cause at all; the other is that we at last have recourse to some ultimate cause that is necessarily existent. . . . But if the whole eternal chain of succession, taken together, is not determined or caused by anything, this is absurd. . . . We must, therefore, have recourse to a . . . Being who carries the reason of his existence in him, and who cannot be supposed not to exist, without an express contradiction."

The holistic, dialectic argument for the existence of God was the following: " . . . Just as two people watch a cup on the table, one sees a cup with a handle, the other must see a cup without a handle if he is looking from the opposite perspective . . . each one of them can only see a part of the truth. Is nothing the ultimate truth? . . . there must be a way to add up all the different perspectives. . . . Such a sum or 'whole' consists of every idiosyncratic perspective, but reveals the truth as a whole. This marvelous 'whole' cannot be designed or found by any individual alone. We must, therefore, have recourse to a necessarily existent Being who is above every idiosyncratic entity . . ."

A majority of Americans preferred Galileo's logical argument against Aristotle's assumption about gravity whereas a majority of Chinese preferred the holistic, dialectical argument. A majority of Americans preferred the "logical" argument about the existence of God over the holistic argument we concocted, whereas a majority of Chinese preferred the holistic argument. My Western scientific colleagues find the Chinese preference for the holistic argument against Aristotle's views to be astonishing, since they regard Galileo's argument as knockdown. So I should note that only 60 percent of Americans preferred Galileo's argument.

What would happen if Easterners and Westerners were confronted with apparently conflicting propositions? The logical approach would seem to require rejecting one of the propositions in favor of the other in order to avoid a possible contradiction. The dialectical approach would favor finding some truth in both, in a search for the Middle Way. In order to examine this question, Peng and I asked undergraduates at the University of Michigan and Beijing University to read what we described as summaries of the results of several social science studies. There were five different topics altogether and we asked participants either to read about a study reporting a particular finding, a study strongly implying something quite different, or both. The opposing studies did not necessarily contradict each other in a logical sense, but at least had the character that, if one was true, then the other would seem to be quite unlikely to be true. The pair of statements below was typical of the more obviously contradictory ones.

Statement A: "A survey found that older inmates are more likely to be ones who are serving long sentences because they have committed severely violent crimes. The authors concluded that they should be held in prison even in the case of a prison population crisis."

Statement B: "A report on the prison overcrowding issue suggests that older inmates are less likely to commit new crimes. Therefore, if there is a prison population crisis, they should be released first."

The pair of statements below was typical of those that were not contradictory in a logical sense.

Statement A: "A social psychologist studied young adults and asserted that those who feel close to their families have more satisfying social relationships."

Statement B: "A developmental psychologist studied adolescent children and asserted that those children who were less dependent on their parents and had weaker family ties were generally more mature."

If it were really the case that young people who feel close to their families have more satisfying social relationships, then you would not be likely to think that it is also the case that adolescents who have weaker family ties are more mature, though admittedly this would entail no logical contradiction.

Participants rated how believable the statements were. Each pair of statements was composed of one that was more plausible (to both Chinese and Americans) than the other, which we know by looking at the ratings of participants who read only one statement or the other.

What inferences should the participants have made? That seems pretty clear. The participants who were exposed to two propositions that are apparently contradictory ought to have believed in each of them less than those who knew about only one. This should be particularly true for less plausible propositions that are countered by more plausible ones. But neither the Americans nor the Chinese behaved that way. The Chinese who saw both propositions reported about equal belief in both. They properly rated the more plausible proposition as less believable if they saw it contradicted than if they didn't. But the Chinese rated the less plausible proposition as more believable if they saw it contradicted than if they didn't. This inappropriate inference would be the consequence of feeling it necessary to find the truth in each of two contradictory propositions. The Americans, instead of converging in their belief in the two propositions, actually diverged, believing the more plausible proposition more if they saw it contradicted than if they didn't. This seems the likely result of feeling it necessary to decide which of two conflicting propositions is correct. But it's pretty dubious inferential practice to believe something more if it's contradicted than if it isn't. My guess is that the Americans behaved the way they did because they are good at generating counterarguments—a skill that comes from a lifetime of doing just that. When confronted with a weak argument against a proposition they are inclined to believe, they have no trouble in shooting it down. The problem is that the ease with which they generate counterarguments may serve to bolster their belief in a proposition that ought to seem shakier if it is contradicted than if it is not. There is evidence in fact that Americans do tend to generate more counterarguments than Chinese do. In effect, Americans may not know their own strength, failing to understand how easy it is for them to attack an argument they find implausible.

The American tendency to avoid contradiction seems related to the long-standing Western inclination to search for principles that will justify beliefs. If I can show that some principle is guiding my beliefs, then I can demonstrate that, any appearances to the contrary notwithstanding, my beliefs are consistent with one another. Westerners' need to demonstrate that their beliefs are guided by principles appears to apply also for actual choices. Organizational psychologists Briley, Morris, and Simonson studied the consumer choices of European Americans and people from Hong Kong. All choices were among a triad of objects—computers, for example—that differed on two dimensions. "IBM" was superior to both "Sony" and "Apple" on one dimension and "Apple" was superior to both "IBM" and "Sony" on the other dimension. Sony was always intermediate between IBM and Apple on both dimensions. On average, across the range of choices, Americans and East Asians in a control condition were about equally likely to choose intermediate Sony. In an experimental condition, Briley and colleagues had participants give reasons for their choice, anticipating that this would prompt Americans to look for a rule that would justify a given choice (e.g., "RAM is more important than hard drive space"), but would prompt people of Asian culture to seek a compromise ("Both RAM and hard drive space are important"). When asked to justify their choices, Americans moved to a preference for one of the extreme objects whose choice could be justified with reference to a simple rule, whereas Asian participants moved to a greater preference for the compromise object. Participants gave justifications that were consistent with their choices: Americans were more likely to give rule-based justifications and Chinese were more likely to give compromise-based justifications.

So there is ample evidence to indicate that Easterners are not concerned with contradiction in the same way that Westerners are. They have a greater preference for compromise solutions and for holistic arguments and they are more willing to endorse both of two apparently contradictory arguments. When asked to justify their choices, they seem to move to a compromise, Middle Way stance instead of referring to a dominating principle. The greater adherence to the principle of noncontradiction on the part of Americans seems to produce no guarantee against questionable inferences. On the contrary, Americans' contradiction phobia may sometimes cause them to become more extreme in their judgments under conditions in which the evidence indicates they should become less extreme. This tendency mirrors complaints about hyperlogical Western habits of mind often expressed by philosophers and social critics of both East and West.

Hokum, Emotion, and Math

One of the most reliable phenomena of social psychology is the Barnum effect, named after the circus owner who gave us the expression, There's a sucker born every minute. If you want to make someone, anyone, think that you have remarkable insight into their character, you can just tell them something like the following: "Although generally you have an upbeat personality, sometimes you find yourself blue—without always having a clear idea why. While most people think you are reasonably outgoing, the truth is you are rather shy at the core . . ."

Most everyone thinks they are fairly upbeat but get sad at times, that they seem sociable but are really rather shy. What people don't realize is how common these self-perceptions are and so they feel that the psychologist or fortune-teller, as the case may be, has looked deep into their soul and found truth. Incheol Choi argued that this is made easier if people don't recognize the near contradictions that are carefully built into these phony personality descriptions that lend them plausibility, whatever the person thinks about his personality. If so, then East Asians could be expected to be more susceptible to the Barnum effect, accepting apparently opposing personality descriptions of themselves. To test this, Choi asked Koreans and Americans to rate their personalities on a number of scales. Different scales were designed to tap what most people would say are opposite traits. Choi asked participants to rate how rude they were and, in another part of the questionnaire, how polite they were. Koreans who said they were more polite than others were likely to say that they were about as rude as others. Americans who said they were more polite said they were less rude, or, if they said they were less polite, tended to say they were more rude. A red flag apparently went up for Americans indicating possible contradiction, but was less likely to do so for Koreans.

In an even more striking demonstration of inconsistency, Choi gave Korean and American participants a large number of statements that were literal or near-literal opposites of each other.

- A person's character is his destiny; or
  A person's character is not his destiny.
- The more one knows, the more one believes; or
  The more one knows, the less one believes.

Choi gave some participants one of the opposed pair of propositions and some participants the other. If the Americans given the first statement of the pair tended to agree with it, the Americans given the other statement tended to disagree with it. But this was not necessarily true for Koreans, who were likely to agree with whichever statement of the pair they saw.

There is a poem by William Butler Yeats called "Lapis Lazuli." It describes a gemstone with a carving showing a pair of elderly Chinese men under a pagoda roof on a mountainside.

> *There, on the mountain and the sky,*
> *On all the tragic scene they stare.*
> *One asks for mournful melodies;*
> *Accomplished fingers begin to play.*
> *Their eyes mid many wrinkles, their eyes,*
> *Their ancient, glittering eyes, are gay.*

It may be that Yeats was right to make his point with people who were Chinese, because there is evidence that the simultaneous experience of conflicting emotions is more common for Easterners than for Westerners. Kaiping Peng and his colleagues asked Japanese and American participants to look at faces and to indicate what kinds of emotions they expressed. For Americans, faces were happy or sad, angry or frightened. The more they reported seeing positive emotions, the less they reported seeing negative emotions. (Western) common sense and lots of data collected over the years by psychologists suggest things could scarcely be otherwise. But indeed they were otherwise for the Japanese participants. They were quite likely to report seeing both positive and negative emotions in the same face.

East Asians also seem to have no trouble accepting

apparent contradictions in their own emotions. Organizational psychologists Richard Bagozzi, Nancy Wong, and Youjae Yi asked Chinese, Korean, and American participants to rate their emotional states at the moment and their emotional states in general. American participants tended to report experiencing uniformly positive emotions or uniformly negative ones. But for Chinese and Korean respondents there was little relationship between the intensity of positive emotions they reported, both now and in general, and the intensity of negative emotions they reported. Reporting strong positive emotions was fully compatible with expressing strong negative emotions. Confucius was apparently speaking for at least a very large fraction of the world's people when he said, "When a person feels happiest, he will inevitably feel sad at the same time."

I am sometimes accused of a contradiction myself. Why do nonlogical Asians tend to do so much better in math and science than Americans? How can this be if East Asians have trouble with logic? There are several answers to this question.

First, it should be noted that we don't actually find East Asians to have trouble with formal logic, we just find them to be less likely to use it in everyday situations where experience or desire conflicts with it. Second, Eastern lack of concern about contradiction and emphasis on the Middle Way undoubtedly does result in logical errors, but Western contradiction phobia can also produce logical errors.

The Eastern reputation for math skills is really quite

recent. Traditional Chinese and Japanese culture emphasized literature, the arts, and music as the proper pursuits of the educated person. In research with young and elderly Chinese and Americans, we and others find that only the young Chinese outperform their American counterparts. Comparably schooled older Chinese and Americans perform similarly in math.

Asian math education is better and Asian students work harder. Teacher training in the East continues throughout the teacher's career; teachers have to spend much less time teaching than their American counterparts; and the techniques in common use are superior to those found in America. (Asian math-education superiority to Europe in these respects is less marked.) Both in America and in Asia, children of East Asian background work much harder on math and science than European Americans. The difference in how hard children work at math is likely due at least in part to the greater Western tendency to believe that behavior is the result of fixed traits. Americans are inclined to believe that skills are qualities you do or don't have, so there's not much point in trying to make a silk purse out of a sow's ear. Asians tend to believe that everyone, under the right circumstances and with enough hard work, can learn to do math.

In short, Asian superiority in math and science is paradoxical, but scarcely contradictory!

I have presented a large amount of evidence to the effect that Easterners and Westerners differ in fundamental assumptions about the nature of the world, in the focus of attention, in the skills necessary to perceive relationships and to discern objects in a complex environment, in the character of causal attribution, in the tendency to organize the world categorically or relationally, and in the inclination to use rules, including the rules of formal logic. Two major questions arise in light of these contentions. Does it matter? Is it going to continue? Chapter 8 addresses the former question and the epilogue addresses the latter.

CHAPTER 8

AND IF THE NATURE OF THOUGHT IS NOT EVERY-WHERE THE SAME?

Differences between Easterners and Westerners have been found in virtually every study we have undertaken and they are usually large. Most of the time, in fact, Easterners and Westerners were found to behave in ways that were qualitatively distinct. Americans on average found it harder to detect changes in the background of scenes and Japanese found it harder to detect changes in objects in the foreground. Americans in general failed to recognize the role of situational constraints on a speaker's behavior whereas Koreans were able to. The majority of Koreans judged an object to be more similar to a group with which it shared a close family resemblance, whereas an even greater majority of Americans judged the object to be more similar to a group to which it could be assigned by a deterministic rule. When confronted with two apparently contradictory propositions, Americans tended to polarize their beliefs whereas Chinese moved toward equal acceptance of the two propositions. When shown a thing, Japanese are twice as likely to regard it as a substance than as an object and Americans are twice as likely to regard it as an object than as a substance. And so on.

The lesson of the qualitative differences for psychologists is that, had the experiments in question been done just with Westerners, they would have come up with conclusions about perceptual and cognitive processes that are not by any means general. And in fact just such mistaken conclusions about universality have been mistakenly reached for many of the processes reported on in this book. It seems clear that we need a reconsideration of which perceptual and reasoning processes are basic and which are subject to substantial variation from one human group to another. The fault lines are going to lie deeper, and in different locations, than has been suspected up till now.

Does It Matter?

But the results reported in the body of the book are based mostly on laboratory tests: Why should we assume the findings are anything more than hothouse plants that have no counterpart in real-world thought or behavior?

The question is a fair one and it will be instructive to attempt to answer it. There are in fact many domains of life in which Easterners and Westerners think and behave quite differently and these differences are well understood in terms of our claims about holistic vs. analytic thought.

Medicine Medicine in the West retains the analytic, object-oriented, and interventionist approaches that were common thousands of years ago: Find the offending part or humour and remove or alter it. Medicine in the East is far more holistic and has never until modern times been in the least inclined toward surgery or other heroic interventions. Health is the result of a balance of favorable forces in the body; illness is due to a complex interaction of forces that must be met by equally complex, usually natural, mostly herbalist remedies and preventives. Dissection of bodies into their component parts was practiced by the ancient Greeks and, with a hiatus during the Middle Ages, has been practiced in the West for the last five hundred years, as well. Dissection was not introduced—from the West, of course—to Eastern medicine until the nineteenth century.

Law Contemplate the following equation: First, we define a society's preference for lawyers over engineers as a ratio:

$$\frac{\text{Number of lawyers in the society}}{\text{Number of engineers in the society}}$$

Next, we define a ratio of such ratios as two countries' relative preferences for lawyers over engineers.

$$\frac{\text{Number of lawyers/engineers in society A}}{\text{Number of lawyers/engineers in society B}}$$

The number we get when we divide the lawyer-preference ratio of the United States by the lawyer-preference ratio of Japan is forty-one!

Those lawyers in the U.S. are put to good use. Conflict between individuals in Western countries is handled to a substantial degree by legal confrontations, whereas it is much more likely to be handled in the East by intermediaries. In the West, the goal is satisfaction of a principle of justice and the presumption going into the arena of conflict resolution is typically that there is a right and a wrong and there will be a winner and a loser. The goal in Eastern conflict resolution is more likely to be hostility reduction and compromise is assumed to be the likely result. Westerners call on universal principles of justice to push their goals and judges and juries feel obligated to make decisions that they believe would hold for everyone in approximately similar circumstances. In contrast, in the East, flexibility and broad attention to particular circumstances of the case are the earmarks of wise conflict resolution. As a citizen of prerevolutionary China put it: " . . . A Chinese judge cannot think of law as an abstract entity, but as a flexible quantity as it should be personally applied to Colonel Huang or Major Li. Accordingly, any law which is not personal enough to respond to the personality of Colonel Huang or Major Li is inhuman and therefore no law at all. Chinese justice is an art, not a science."

Debate Decision processes in Japanese boardrooms and executive councils are designed to avoid conflict and dissonance. Meetings are often little more than a ratification of consensus achieved by the leader beforehand. Japanese managers tend to deal with conflict with other managers by simple avoidance of the situation, whereas Americans are far more likely than Japanese to attempt persuasion. What is intrusive and dangerous in the East is considered a means for getting at the truth in the West. Westerners place an almost religious faith in the free marketplace of ideas. Bad ideas are no threat, at least over the long run, because they will be seen for what they are if they can be discussed in public. There has never been such an assumption in the East and there is not today.

Science In the decade of the nineties, scientists living in the United States produced forty-four Nobel Prizes and the Japanese produced just one, despite the fact that Japanese funding for science is fully half that of the U.S. West Germany, which spends half as much on science as Japan, has produced five Nobel Prize winners. And France, with far less funding even than Germany, has produced three. The relatively slight accomplishments of Japanese science can be chalked up partially to the Confucian respect for elders that funnels support to mediocre older scientists instead of more talented younger ones. But some Japanese scientists attribute the deficit in part to the absence of debate and intellectual confrontation. Peer review and criticism are rare in Japan, where such things are considered rude and where there is not widespread acceptance of their role in clarifying and advancing thought about scientific matters. As one Japanese scientist put it: "I worked at the Carnegie Institution in Washington and I knew two eminent scientists who were good friends, but once it came to their work, they would have severe debates, even in the journals. That kind of thing happens in the United States, but in Japan, never."

Rhetoric The resistance to debate is not merely a social or ideological one, nor is it limited to purely quantitative outcomes, such as the number of scientific papers produced. The reluctance extends to the very nature of communication and rhetoric. Western rhetoric, which provides the underlying structure for everything from scientific reports to policy position papers, usually has some variation of the following form:

- background;
- problem;
- hypothesis or proposed proposition;
- means of testing;
- evidence;
- arguments as to what the evidence means;
- refutation of possible counterarguments; and
- conclusion and recommendations.

Most Westerners I speak to about this format take it for granted that it is universal: How else could one communicate findings and recommendations briskly and convincingly or even think clearly about what one is doing? The truth is, however, that this linear rhetoric form is not at all common in the East. For my own Asian students, I find that the linear rhetoric form is the last crucial thing they learn on their road to becoming fully functioning social scientists.

Contracts To the Western mind, once a bargain is struck, it shouldn't be modified; a deal is a deal. For Easterners, agreements are often regarded as tentatively agreed-upon guides for the future. These opposing views have often caused conflict between Easterners and Westerners. Recall the bitterness between Japanese and Australian businesspeople over Australia's refusal to renegotiate a contract for sugar when the price dropped radically on the world market. The Japanese were not being hypocritical or purely self-serving. Japanese suppliers take such matters under consideration with their own customers. If it snows in Tokyo, film distributors are likely to compensate theater owners for their diminished audiences. As business professors Hampden-Turner and Trompenaars note, "Looked at analytically on an item-by-item basis, [such accommodating behavior] is not cost-effective. But looked at as strengthening the relationship between customer and supplier, it makes very good sense." In short, the Japanese take a holistic view of the business relationship, including its context over time.

International Relations An international conflict influenced by differing conceptions of causality occurred between China and the United States when a Chinese fighter plane collided with an American surveillance plane and the surveillance plane was forced to land on a Chinese island without receiving permission from the ground. The Chinese held captive the crew of the surveillance plane, demanding an apology for the incident from the U.S. The Americans, asserting that the cause of the accident was the recklessness of the fighter pilot, refused. Political scientist Peter Hays Gries and social psychologist Kaiping Peng have observed that, to the Chinese, an insistence that there was such a thing as *the* cause of the accident was hopelessly limited in its perspective. Relevant to the accident were a host of considerations, including the fact that the U.S. was, after all, spying on China, there was a history of interaction between the particular surveillance plane and the particular fighter, and so on. Given the complexity and ambiguity of causality—taken for granted by the Chinese to be the case in this instance as in all others—the very least the United States could do would be to express its regrets that the incident occurred. The presumed ambiguity of causality may lie behind Eastern insistence on apology for any action that results in harm to someone else, no matter how unintentionally and indirectly (and the readiness of Japanese managers to resign when matters over which they could not possibly have had control go awry). Ultimately, the "regret" formula was the one that China and the U.S. hit upon to resolve the impasse, but it is not likely that many people on either side understood the role played in the conflict by the differing conceptions of causality that Gries and Peng identified.

Human Rights Westerners seem inclined to believe there is only one kind of relation between the individual and the state that is appropriate. Individuals are separate units and they enter into a social contract with one another and with the state that entails certain rights, freedoms, and obligations. But most peoples, including East Asians, view societies not as aggregates of individuals but as molecules, or organisms. As a consequence, there is little or no conception of rights that inhere in the individual. For the Chinese, any conception of rights is based on a part-whole as opposed to a one-many conception of society. To the extent that the individual has rights, they constitute the individual's "share" of the total rights. When Westerners see East Asians treating people as if they had no rights as individuals, they tend to be able to view this only in moral terms. Whatever the moral appropriateness of the behavior of East Asian officials—and I share with most Westerners the view that there is such a thing as individual human rights and that they sometimes are violated in East Asia—it is important to understand that to behave differently would require not just a different moral code, but a different conception of the nature of the individual. A different conception of the individual would in turn rest on an inclination to think about the world in terms of individual units rather than continuous substances at the most basic metaphysical level.

It is also important to recognize that East Asians and other interdependent peoples have their own moral objections to Western behavior. When East Asian students become comfortable enough to speak out in Western classrooms, they will often express bewilderment at how much disorder, crime, and exposure to violent and sexually explicit images in the media Westerners are willing to tolerate in the name of freedom. They perceive these issues as entailing human rights because rights are perceived as inhering in the collectivity rather than the individual.

Religion Some of the many religious differences can be understood in terms of the "right/wrong" mentality of the West in contrast to the "both/and" orientation of the East. Eastern religions are characterized by tolerance and interpenetration of religious ideas. One can be a Confucian, a Buddhist, *and* a Christian in Korea and Japan (and in China prior to the revolution). Religious wars in the East have been relatively rare, whereas they have been endemic in the West for hundreds of years: Monotheism often carries with it the insistence that everyone accede to the same notion of God. It could be argued that the Greeks should be held blameless in this (after all, they had many gods and didn't much care which ones any particular individual favored), and perhaps this is true. It is the Abrahamic religions that have been so inclined toward religious warfare. On the other hand, it has been claimed that Christianity is the only religion that finds it necessary to have a theology specifying essential aspects of God and that this insistence on categorization and abstraction is traceable to the Greeks.

Cycles and recurrences are an integral part of many Eastern religions but are less common in the West. Rebirth is part of some Eastern religions but rare in Western ones. Sin is understood to be a chronic condition and can be atoned for in many Eastern religions (as well as in Catholicism to a degree). But sin is hard to atone for or literally ineradicable in the Protestant tradition. You might say that as one moves West from India, the number of possible states after death lowers drastically—from the near infinity of reincarnations of Hinduism and Buddhism to the multiple levels of Catholic purgatory and circles of hell to the binary possibility of the Calvinist.

Finally, it should be recalled that much of the evidence discussed in this book is drawn from everyday life problem solving. Japanese managers start at the bottom of their companies and are rotated among divisions frequently so as to maintain an overview of their companies' activities. Buildings in China, even skyscrapers in Hong Kong, are built only after an exhaustive survey by *feng shui* experts who examine every conceivable ecological, topological, climatologic, and geometric feature of landscape and proposed building simultaneously and in relation to one another. It is Westerners, and Americans in particular, who pioneered the atomistic, interchangeable, uniform, modular approach to manufacturing and merchandising. And so on. My claim is not that the cognitive differences we find in the laboratory cause the differences in attitudes, values, and behaviors, but that the cognitive differences are inseparable from the social and motivational ones. People hold the beliefs they do because of the way they think and they think the way they do because of the nature of the societies they live in.

How Should People Think?

Early in the twentieth century, philosophers and psychologists effected a division of labor. Psychologists were given the descriptive task of finding out how people thought and behaved. Philosophers were assigned the prescriptive job of telling people how they ought to think and behave. Sometimes, though not as often as might have been advisable, philosophers have looked to the work of psychologists to find out what people actually do. But even if philosophers had been paying close attention to the work of psychologists, they would have found little to disabuse them of their convictions about universality. I believe the work reported here will have that effect on psychologists and consequently on philosophers, as well.

To see how philosophy might be affected by demonstrations of nonuniversality, consider the riddle of induction, as introduced by David Hume in the eighteenth century. How are we justified, he asked, in assuming that the future will be like the past, that the food that nourished us today will nourish us tomorrow. There can be no question of a deductive solution to the problem. "This food nourished me today; therefore, it will nourish me tomorrow" has only a probabilistic status; it lacks the certainty that is required of a syllogism.

The philosopher Nelson Goodman proposed that the solution to the riddle of induction is to seek *reflective equilibrium* between rules for inductive inferences and the specific inferences that we in fact make. This is what we do with deductive rules: We would abandon any deductive rule that required us to sanction inferences that we found unacceptable, and would reject any conclusion that was prohibited by a rule we were unwilling to give up. But suppose there are cultures that don't reason as "we" do, and moreover, don't even endorse the same principles of reasoning that we do? Philosopher Stephen Stich has observed that this makes a shambles of the reflective equilibrium principle. If we don't agree about whether an inference is justified or not, we can't use the principle as a guide to correct thinking—just an expression of personal preference. One solution is just to say that we're justified in our inferences and they're justified in theirs—even if their inferences are completely different from ours. This position of extreme relativism is an easy one to take, but no one really believes it. If you tell me that you believe that both of two virtually contradictory propositions are correct, I may politely say that I'm sure you're right *for you* but I'm right *for me*. Is either of us convinced? Probably not.

But I'm not willing to lie in this bed of relativism I've helped to make. On the contrary, I find that Asian patterns of reasoning cast valuable light on some of the reasoning errors of Westerners and I believe the same mirror can be profitably reversed to look at Eastern thought.

I will focus on just a few Western habits of thought that seem particularly illuminated by contrasting them with Eastern patterns of thought.

Formalism There is enormous power in the formal, logical approach of Western thought. Science and mathematics obviously rely on it, though just how much is a matter of dispute. Francis Bacon wrote that "logic is useless; it is creation that is science." And Bertrand Russell expressed the view that the syllogisms of the twelfth-century monks were as sterile as they were. Though I'm inclined to agree, this is a puzzling statement coming from someone who believed that all human problems could be solved by logic, but could apply only formal logic to real-world questions. In my view, this rendered his analysis of political and social questions naive. The chief cause of his problem was the insistence on separation of form and content, so that reasoning could be carried out using logical principles on the form alone. This is a Western ailment. As the philosopher S. H. Liu says, "Chinese are too rational to separate form from content."

A second problem for Russell was that, like most Westerners, he was largely lacking in what may be called the "reasoning schemas" of dialecticism. Many such schemas were identified (without using the term "dialecticism") by developmental psychologists Klaus Riegel and Michael Basseches. These psychologists disagreed with

Jean Piaget's view that most reasoning was carried out by means of so-called formal operations, or logical principles, which were in place by adolescence. In their view, most high-level reasoning was carried out by means of postformal operations—reasoning schemas that are more complex and more tied to specific thought content than are logical rules. They were termed "postformal" because they were assumed to develop primarily after the formal operations were complete. Both Riegel and Basseches believed that progress in development of postformal operations continues throughout the lifespan. Some examples from Basseches's work include the following.

- The concept of movement from thesis to antithesis to synthesis.
- The ability to understand events or situations as moments in the development of a process.
- The recognition of the possibility of qualitative change as a result of quantitative change.
- The ability to take a stance of contextual relativism.
- The recognition of the value of multiple perspectives on a problem.
- The recognition of the pitfalls of formalism based on the interdependence of form and content.
- The ability to understand the concept of two-way reciprocal relationships.
- The ability to understand the concept of self-transforming systems.
- The ability to conceive of systems in terms of their equilibrium.

Oddly, neither Riegel nor Basseches seems to have made the connection in print between their notions about postformal operations and the dialectical aspects of Eastern thought, though it seems highly unlikely that they were unaware of the similarities. In fact, it is probable that they drew on Eastern ideas for developing the schemas.

The two Western vices of separation of form and content and the insistence on logical approaches often operate together to produce a lot of academic nonsense. There are plenty of examples from my field of psychology to point to. In particular, a great deal of formal modeling of psychological phenomena—most that I am aware of—fails to elucidate the phenomena it purports to. The joy lies in modeling for its own sake, not in making sense of behavior. Economist friends have told me that the macho thing in economics is to pick some implausible principle and then derive as many phenomena as possible from it.

Two-valued Logic The binary, "either/or" approach to the evaluation of propositions characteristic of the West has been lamented by many Western thinkers, but the problems are easier to see from the standpoint of the "both/and" approach of the East. For example, the Western insistence that a behavior have *a* cause, rather than a number of causes, results in people seeing behavior *either* as intrinsically caused or extrinsically caused, but not both. So someone can act out of generosity or to satisfy some self-serving motive, but not for both types of reasons. Adam Smith wrote from this perspective in his famous defense of capitalism: "It is not that he cares for you, the customer, that the brewer, the baker, or the butcher provides for your dinner, but because he cares for himself."

But on reflection, why not both motives? Surely many merchants are primarily in business to feed their own families but also like the fact that they are helping to feed others, as well. This was recognized by Smith himself but has been ignored or unappreciated by many of his followers.

There is a cynicism about the motives of politicians that is characteristic of Americans which, however healthy it might be for maintaining personal freedoms, is likely to produce some incorrect assessments. Neither Lyndon Johnson nor Richard Nixon is among my favorite politicians, but both were widely seen as having acted for political gain when they did things that in fact they believed would lead to serious loss. Johnson was seen by many as trying to enhance his political capital by fighting for Kennedy's civil rights bills, but in fact he knew—better than Kennedy could have—that he was signing over the South to the Republican Party for a generation. Nixon was thought by many to be seeking personal political gain by the opening to China when in fact he and many of his aides feared it would be an extremely unpopular move.

There is a bit of evidence that Westerners may be more susceptible to this "single-motive fallacy" than other people. Developmental psychologists Joan Miller and David Bersoff told American and East Indian children about cases in which one person helped another person. In some instances, the helper expected reciprocation and in other cases did not. The Indian children assumed that the helper was intrinsically eager to help, regardless of expectations about reciprocation. The American children believed there was an intrinsic motive to help only if there was no expectation of reciprocation.

The Fundamental Attribution Error One of social psychology's most important and best demonstrated phenomena is the fundamental attribution error—the tendency to assume that the behavior of another person has been produced by personality traits or abilities and to slight important situational factors. Critics have sometimes held that this tendency doesn't constitute an error at all. But East Asians are less susceptible to the error than Americans in some cases and the error is more readily corrected for them when the situation is highlighted in some way. The critic can't have it both ways. Either Westerners are wrong in those cases when they ignore the implications of the situation or Asians are wrong when they take them into consideration. The more plausible position, especially in light of the data showing that Americans are prone to attend only to salient objects and to ignore contexts, is to say that it is the Americans who are wrong and the Asians who are right in these cases.

Research on the fundamental attribution error has philosophical implications beyond the epistemological. The work is also important for ethics, a point emphasized by philosophers including John Doris, Gilbert Harman, and Peter Vranas, as well as by many psychologists. They note that Aristotle's ethics, which has played a large role in the history of Western philosophy, is similar to his physics. People, like objects, behave as they do because of their properties—virtues or vices in the case of the ethically relevant behavior of people. Aristotle's "virtue ethics" is more consistent with lay Western thought about moral behavior than with Eastern beliefs. Aristotle's system encourages you to assume that people are incorrigible or to take the stance that behavior must be altered by changing people's attributes—a difficult thing to do at best and counterproductive at worst. If you want to get people to behave as you (and often they) believe they ought, an easier route is to encourage them to seek out situations that will bring out the best behavior in them and to shun those that will encourage bad behavior. Such an approach to encouraging ethical behavior is more obvious from an Eastern viewpoint than from a Western viewpoint.

Turnabout is fair play, and it is also possible to use Western principles as a platform for criticizing Eastern thought. A sketch of what that enterprise might look like follows.

Contradiction The heuristic "there's truth on both sides" may very well be a good one to use as a first approach to understanding any apparent contradiction. It may also be a good place to end up much of the time. It is not an algorithm best followed relentlessly, however. Sometimes one proposition has all or most of the truth on its side and the other has little or none. We have seen that Easterners are more willing to grant credence to each of two propositions that bear a contradictory relationship to each other than Americans are, and that they can be led into the serious error of believing a given proposition more when they see it contradicted by a more plausible proposition than when they merely see it by itself. This is almost impossible to defend on logical grounds but can readily be seen as the result of an insistence on finding the Middle Way. Incheol Choi maintains that the relative insensitivity of Easterners to contradiction makes it less likely that they will have sufficient curiosity to become scientists. Whether this is a good or bad thing is a matter of preference, but it is certainly relevant that the people who run Eastern societies at the moment happen to want to be able to produce scientists.

Debate and Rhetoric I share the Western conviction about the efficacy of debate for bringing out the truth or, at any rate, for keeping on the table hypotheses that may be useful. Western debate style, and the mental habits it encourages, are important for keeping societies open and open-minded. Debate also goes hand in hand with standard hypothesis-evidence-conclusion rhetoric, which science and mathematics rely on heavily. Earlier I quoted physicist Alan Cromer to the effect that "a geometric proof is the ultimate rhetorical form." Statistician and psychologist Robert Abelson has written a lovely book describing statistics essentially as rhetoric. I believe the metaphors are deep and correct.

Complexity A Western thinker has said that "if the universe is pretzel-shaped, then we must have pretzel-shaped hypotheses." True enough, but if we start with a pretzel-shaped hypothesis, the universe had better be pretzel-shaped or there's no chance we'll find out just what shape it *is*. For any shape other than a pretzel, you're better off starting with a straight line and modifying it as it becomes clear that the linear hypothesis is too simple. Asians are surely right in their belief that the world is a complicated place and it may be right to approach everyday life with this stance. In science, though, you get closer to the truth more quickly by riding roughshod over complexity than by welcoming onboard every conceivably relevant factor.

Of course, prescriptive observations like those in this

section only make sense if we think that people's habits of mind can be altered readily. Can they be?

TEACHING AND TESTING

Should educators seek to give other cultures' skills to its children or should they focus on what is defined as important in their own culture?

Americans are so used to hearing about the educational successes of Asians and Asian Americans both in Asia and in the U.S. that it comes as a shock to hear about children of U.S.-based Japanese businessmen who are labeled "learning disabled" in American schools and put back. Their inability to perform causal analysis—for example, in history classes—in the most rudimentary way expected of American children leads to the belief that they are cognitively impaired.

Causal analytic skills are not the only respect in which Asians are sometimes held deficient by American educators. Debate is an important educational tool for learning analytic thinking skills and for forcing self-conscious reflection on the validity of one's ideas. This view is shared increasingly by non-Westerners. Debate training is becoming a minor American export industry, with young people from all over the world, but especially Asia, coming to debate camps in the U.S.

A few years ago, Heejung Kim, a graduate student from Korea studying psychology at Stanford, became exasperated with the constant demand from her American instructors that she speak up in class. She was told repeat-

edly that failure to speak up could be taken as an indica-
tion of failure to fully understand the material and that, in
any case, speaking up and hearing the reactions of the
instructor and classmates would help her to understand it
better. Kim didn't believe it. Instead, she felt that she and
her fellow Asian and Asian American students would not
benefit from speaking because their fundamental way of
understanding the material was not verbal. There is certainly a long tradition in the East of equating silence rather
than speech with knowledge. As the sixth-century B.C.
sage Lao-tzu said, "He who knows does not speak, he who
speaks does not know." Kim explains the difference by
calling on the distinction made in our work between analytic and holistic thought. Analytic thought, which dis-
sects the world into a limited number of discrete objects
having particular attributes that can be categorized in
clear ways, lends itself to being captured in language.
Holistic thought, which responds to a much wider array of
objects and their relations, and which makes fewer sharp
distinctions among attributes or categories, is less well
suited to linguistic representation.

To test the possibility that Asians and Asian Americans
in fact find it relatively difficult to use language to repre-
sent thought, Kim had people speak out loud as they
solved various kinds of problems. This had no effect on the
performance of European Americans. But the requirement
to speak out loud had very deleterious effects on the performance of Asians and Asian Americans. This work is as
convincing as any in this book about the different nature
of thought for Asians and Westerners and its practical
implications are extremely important. How should one

educate Asians and Asian Americans in American class-rooms? Is it a form of "colonialism" to demand that they perform verbally and share their thoughts with their class-mates? Would it have the effect of undermining the skills that go with a holistic approach to the world? Or is it merely common sense to prepare them for a world in which verbal presentation skills, even if it might be diffi-cult to achieve them, will come in handy?

Two advantages of Asian cognition stand out: (1) the fact that Asians see more of a given scene or context than Westerners do; and (2) the holistic, dialectic, Middle Way approach to problems. Leaving aside for the moment the question of whether one should attempt to teach these skills to Westerners, there are some hints from the work of cognitive psychologists David Meyer and David Kieras that it might be surprisingly easy to open "bottlenecks" in perceptual and perceptual-motor performance. People can be taught to attend to a broader range of different stimuli, and respond to them more quickly and accurately, with only modest amounts of training. The cognitive aspects of holistic, dialectic approaches to reasoning seem to me to be a different matter entirely. They are so embedded in perception, philosophy, and even temperament that it seems doubtful that much in the way of change could be achieved. But I would be delighted to be proved wrong.

An unchallenged assumption of intelligence testing for the past century is that it is possible to test intelligence in a culture-fair way. The experts agree that cultural biases can creep into language-based intelligence tests. Even within a given culture, people of different socioeconomic status have different exposure to words, and certainly across cultures and across languages, comparisons become almost meaningless. But there is consensus that if intelligence is tested without the use of words, it is reasonable to compare people from different cultures.

Have a look at the illustration on page 214 with its many boxes. It shows a problem similar to those found on well-known tests purporting to be culturally unbiased, such as the Cattell Culture-Fair Intelligence Test and Raven's Progressive Matrices Test. The task for the person being tested is to look at the first few objects in the matrix at the top and figure out what the next object, among the six options shown beneath the matrix, should be. Everyone has been exposed to circles and rectangles and triangles, so there would seem to be no question of unfair advantage there. It should just be raw intelligence that is being measured. But viewed in the light of ideas proposed in this book, the test can be seen to play to the strengths of Westerners. It consists of identifying relevant features, deciding how to categorize them, and finding the rule that best accounts for the way the categories are manipulated.

With a research team headed by Denise Park and Trey Hedden at the University of Michigan and Qicheng Jing of the Chinese Institute of Psychology, I tested the intelligence of American and Chinese college students and elderly people in three different ways: by means of speed and memory tests that are correlated with IQ scores (at least in Western populations where the question has been examined); by percentile score for general information in the relevant comparison population (also highly correlated with IQ scores); and by the Cattell Culture-Fair Intelli-

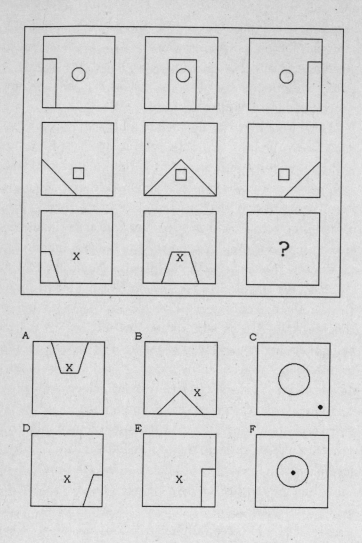

Example of "culture-fair" IQ test item.

gence Test. We equated each of the groups for speed and memory, so that young Americans and Chinese had identical scores on average, as did older Americans and Chinese (the young are much faster and have better memories, so it was not possible to equate across age groups for these variables), and had identical percentile scores for information as well (the elderly, in our samples as is usually the case, had somewhat higher information scores than the young). Despite this matching on two very different measures of intelligence, the Americans, both young and old, scored substantially better on the "culture-fair" test than the Chinese. The difference was very substantial (more than four-fifths of a standard deviation, for readers familiar with statistics). If we took the results of the Cattell Test seriously, and didn't have the other information about abilities, we would have to conclude that Americans were a lot smarter than Chinese (or would if we had any claims to having a random sample of the relevant populations, which we don't).

Now have a look at the illustration on page 216. The person being tested is told to look at the block at the top and produce a "running bird" and a "flying bird" by proper arrangement of the numbered pieces. (To save the reader the trouble of doing this, I've provided the answers at the bottom!) This item looks like it might have been produced by the Educational Testing Service for measuring the spatial relations aptitude of high school seniors. In fact, the problem is more than a thousand years old, having been designed for the purpose of selecting the mandarinate of China. For whatever reason, Chinese and Japanese today teach elementary students how to solve problems like this. In addition, the particular kinds of spatial analysis required to read and write ideographs, and the holistic nature of Asian cultures, seem likely to foster spatial skills. And

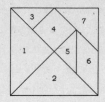

Problem: Make a running bird and a flying bird out of the shapes above

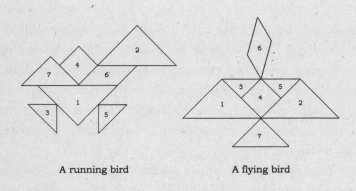

A running bird   A flying bird

"Seven-Coincidence Board."

indeed, Asians and Asian Americans are generally found to outperform European Americans on spatial tasks. (The differences are usually quite large—typically the better part of a standard deviation.) If there were any reason to assume that populations were being sampled randomly (which there isn't), this might encourage some people to contend that East Asians are more intelligent than people of European culture. And indeed it has. Just such an assertion is included among the myriad dubious propositions in the book *The Bell Curve* by Richard Herrnstein and

Charles Murray—together with the assertion that the finding is strong evidence of a genetic basis for the difference, since such spatial tests are obviously culture fair.

Ethnic diversity has been acclaimed for all sorts of reasons, among them that educational and work environments are enriched by having people of different backgrounds. Our work does strongly support the contention that diverse views should be helpful for problem solving. The cognitive orientations and skills of East Asians and people of European cultures are sufficiently different that it seems highly likely that they would complement and enrich one another in any given setting. We would expect that for most problems one would be better off having a mix of people from different cultures than having people who are all from one culture.

Whether such an advantage to diversity will endure depends on whether we are engaged in a worldwide homogenizing process.

THE END OF PSYCHOLOGY OR THE CLASH OF MENTALITIES?

Social scientists in many fields are now debating two very different views of the future. One view, championed by political scientist Francis Fukuyama, assumes convergence of world political and economic systems, and consequently of values, and the other predicts continued difference. Fukuyama has written of "the end of history," meaning that capitalism and democracy have won and that there are no forces on the horizon that can generate interesting events (as in the Chinese curse, May you live in interesting times). The other view, championed by political scientist Samuel Huntington, predicts continued difference. Far from accepting Fukuyama's vision of societal convergence, Huntington has pronounced the world to be on the brink of a "clash of civilizations," with major cultural groups including East Asia, Islam, and the West locked in opposition to one another due to irreconcilable differences in values and worldviews: "In the emerging world of ethnic conflict and civilizational clash, Western belief in the universality of Western culture suffers three problems: it is false, it is immoral, and it is dangerous."

Of course, if economies and governmental forms are to be everywhere the same, this would suggest that the psychological characteristics of peoples will be the same, as well. On the other hand, a clash of civilizations suggests the possibility of a continued divergence in habits of thought. So will the cognitive differences documented in this book turn out to be of mere historical interest? Are they going to be gone in fifty or one hundred years because social systems and values have converged? Will the universalists then turn out to be right, though for the wrong reasons? (Right because everyone will think in the same way, wrong because the reasons for it will not be biological but cultural.) Or will they persist—as they have for thousands of years?

WESTERNIZATION?

Fukuyama's views capture those of many in the West— perhaps especially Americans, who tend to assume that everyone is really an American at heart, or if not, it's only a matter of time until they will be. There is plenty of superficial evidence to back up this belief. People in every country wear jeans and T-shirts and Nike shoes and drink Cokes and listen to American music and watch American movies and TV. (Even France felt it necessary recently to ration the amount of TV of American origin to 25 percent of the total. On the other hand, they've thrown in the language towel and all French elementary schoolchildren will henceforward learn English.) Asian scholars have assured me that higher education in Asia is ever more Western in nature—emphasizing analysis, criticism, logic, and formal approaches to problem solving.

There is some evidence that socialization of children in the East is moving toward the Western pattern. Harold Stevenson and his colleagues monitored the mothers of children in a particular elementary school in Beijing for more than a decade beginning in the mid-eighties, asking them what it was that they wanted for their children. When the study began, the mothers' concerns were for their children's relational skills—their ability to fit in harmoniously with others. Ten years later, the mothers were interested mostly in the same things that Western mothers are: Does my child have the skills and the independence to get ahead in the world?

A few years ago Kaiping Peng, Nancy Wong, and I began to realize that many value surveys were actually showing that Easterners were reporting that they held certain "Western" values more strongly than Westerners did. Indeed, we ourselves found that Beijing University students reported valuing equality, imaginativeness, independence, broadmindedness, and a varied life more than did University of Michigan students, whereas Michigan students reported valuing being self-disciplined and loyal, even having respect for tradition and honoring parents and elders, more than did Beijing students! (My experience as a parent of two University of Michigan students makes me particularly dubious about this last finding.) The odd results are probably partly due to the fact that value checklists, and even attitude scales, are not very good ways of getting at values. When we described scenarios that tacitly pitted values against one another and asked participants how they would behave in those situations, or would prefer for others to behave, we got results that matched the intuitions of Asian and American scholars who study Asia. But if there is any truth to the idea that people tend to become what they are trying to become, or what they say they are, the value surveys may be an augury of the future.

CONTINUED DIVERGENCE?

In Huntington's opinion, the assumption that the world's cultures will be assimilated to those of the West is an illusion bred of myopia and ethnocentrism. The societal differences are sufficiently great that future international conflicts will be more nearly cultural in origin than economic or political as in the past. Islam, the East (especially China), and the West are on divergent cultural paths and the relative influence of the West, because of the economic advances of the Far East and the demographic growth of Islam, is going to decline. The world is not necessarily going to be safe for democracy or free markets.

There is certainly evidence that one can call upon in support of this view.

Japan has had a capitalist economy for more than one

hundred years and capitalism can be expected to promote values of independence, freedom, and rationalism. Yet there are numberless signs that Japan has changed little in many social respects and we find large differences between the way Japanese and Westerners perceive the world and think about it. Capitalism itself has been altered to cohere with Japanese social values. Company loyalty and team spirit, consultative management, and cooperativeness across industries all arose from Japanese social values; many held them to be largely responsible for the "Japanese miracle" of economic development in the post–World War II period. Indeed, it was widely assumed fifteen years ago that the West would have to move toward Japanese forms of management and business practices in order to be able to compete. Of course, Japan's current economic woes are widely attributed to essentially the same social values as its former success. Many Western observers (some of the same ones, in fact!) now regard those values as liabilities resulting in too much reluctance to downsize and too much readiness to make loans to friends in companies having dubious economic prospects.

Japan has had a democratic form of government since shortly after World War II, but its constitution was written for it by Americans and many would say that the government more nearly resembles an oligarchy than a democracy—at least until very recently. And, in any case, it's not clear how long a nation has to be a democracy before one can say it is likely to remain that way, especially when there are serious economic strains.

China, of course, shows little interest in democracy at this point—or, at any rate, it certainly looks like its adherents have an enormous job cut out for them. China's embrace of capitalism is also less than convincing at this point. Korea seems to take more wholeheartedly to free market practices, but democracy is scarcely five years old in that country. And both countries of course remain heavily Eastern in a cognitive sense.

As Huntington has observed, Westerners tend to confuse modernization—defined as industrialization, a more complex occupational structure, increased wealth and social mobility, greater literacy, and urbanization—with Westernization. But societies other than Japan have become modern without becoming very Western. These include Singapore, Taiwan, and, to a lesser degree, Iran. Anyone assuming that modernization can only bring more Westernization should be given pause by the current estimate that by 2007 the most common language used on the Internet will be Chinese and the prediction by some economists that within a few years as much as half the world's international air traffic will involve travel through Pacific Asia.

In short, values continue to diverge and anyone who thinks not is confusing the drinking of Cokes and the building of computers with Westernization.

CONVERGENCE?

But a third view should be considered, which is that the world may be in for convergence rather than continued divergence, but a convergence based not purely on Westernization but also on Easternization and on new cognitive forms based on the blending of social systems and values.

There are certainly indications that the West finds attractions in the East. While the rest of the world drinks Cokes and wears jeans, Westerners are rapidly fusing their cuisines with Eastern ones. Korea's populace is now one-third Christian, but the countless resorts in the Catskill Mountains formerly catering to a middle-class Jewish clientele are rapidly transforming themselves into centers for the study of Buddhism—which is gaining U.S. adherents at a much more rapid rate than mainline Protestantism. Many mainstream Western doctors accept some of the general notions of holistic medicine, even recommending ancient Asian treatments in lieu of modern Western ones for ailments ranging from headache to nausea. More importantly, the need to treat the whole person rather than attack "the" problem has gained wide currency. Millions of Americans, many of them not otherwise trendier than the soccer mom or insurance agent next door, now practice yoga and tai chi. Many Americans who find the traditions of individualism to be alienating look to Eastern forms of community as possible cures for social anomie. Whole industries now practice Japanese-pioneered forms of employer-employee relations. While Easterners learn to emphasize debate in education, Westerners experiment with logical systems that do not require that a proposition be either true or false. Great twentieth-century physicists, such as Nils Bohr, have attributed their progress in quantum mechanics to an appreciation of Eastern ideas. At a time when Western primatologists believed that only the mother-infant bond was an important relationship for chimpanzees, Japanese primatologists were seeing complex interrelationships in stable chimpanzee societies. Initially dismissed, the Japanese view is now the universally accepted one in the field. And, although I have not stressed the point, it should be clear that the ideas in this book owe as much to Eastern thinkers and experimentalists as to Western. I firmly believe that the entry of East Asians into the social sciences is going to transform how we think about human thought and behavior across the board.

If social practices, values, beliefs, and scientific themes are to converge, then we can expect that differences in thought processes would also begin to evaporate. There is in fact evidence that changes in social practices, and even changes in temporary states of social orientation, can change the way people perceive and think.

Recall that many of our studies included Asian Americans. Since they have very different social experiences from those of Asians, we would expect that their perceptions and patterns of thought would resemble those of other Westerners to a substantial degree. And in fact the perceptual patterns and reasoning styles of such participants were always intermediate between those of Asians and European Americans and sometimes were actually indistinguishable from those of European Americans.

Other work suggesting that cognitive modifiability is possible comes from the study of genuinely bicultural people. Evidence suggests that such people do not merely have values and beliefs that are intermediate between two cultures, but that their cognitive processes can be intermediate, as well—or at least that they can alternate between forms of reasoning characteristic of one culture versus another. Recall the studies on causal perception showing that people from Hong Kong can be "primed" by showing them Western symbols such as Mickey Mouse and the U.S. Capitol, and that this prompts them to answer causal questions in a more Western fashion than if they are primed by Eastern symbols, such as temples and dragons. Similarly, Asian Americans answered questions about physical causality in a more Western fashion if they first were asked to recall an experience that made their identity as an American apparent to them than if they recalled an experience that made salient their identity as an Asian.

Shinobu Kitayama and his colleagues found evidence that cognitive processes could be modified even after relatively limited amounts of time spent in another culture. In a particularly elegant demonstration, they presented Japanese and American participants with several examples of a line drawn within a square. Then they were taken to another part of the room and shown a square of a different size than the one they had just seen. They were asked to draw a line inside the square either of the same length they had just seen or that was proportionally the same. Americans were more accurate in drawing a line that was the same absolute length, showing that they were more capable of ignoring the context. Japanese were more accurate in drawing a line that was the same relative length, showing that they were more capable of relating object to context. Then Kitayama and colleagues went a step further and looked at the behavior of Americans who had been living in Japan for a period of time (usually a few months) and Japanese who had been living in America for a period of time (usually a few years). Americans living in Japan were shifted in a decidedly Japanese direction.

Japanese living in America were virtually indistinguishable from native Americans. The study does not really prove that time in another culture produces such dramatic changes in behavior; other interpretations are viable, including the possibility that people who go to live in another culture are very much like them before they ever get there. But the results are strongly suggestive that cognitive processes can be modified by dint of merely living for a time in another culture.

In a sense, we are all "bicultural" with respect to social constraints and social interest. Our awareness of connections with other people, as well as how much we want to associate with other people, varies from time to time. Are these fluctuating differences in the relevance of other people associated with differences in perception and thought? Social psychologist Ulrich Kühnen and his colleagues have conducted some remarkable studies that indicate that simple laboratory manipulations of social orientation have an effect on the way we think. For example, they tried to "prime" an interdependent, collectivist orientation by having their participants read a paragraph and circle all first-person plural pronouns (we, us, our) and tried to prime an independent, individualist orientation by having them circle all first-person singular pronouns (I, me, mine). They found that interdependence-primed participants were more field dependent than were independence-primed participants as indicated by the Embedded Figures Test; that is, they found it harder to recognize a simple figure that was enmeshed in a more complicated context. Kühnen and Daphna Oyserman, using the same manipulation, found that people were able to remember the contexts in which they had seen particular objects—the result of perceptual "binding" of object and field—better after interdependence priming than after independence priming.

Thus we all function in some respects more like Easterners some of the time and more like Westerners some of the time. A shift in characteristic social practices could therefore be expected to produce a shift in typical patterns of perception and thought.

So I believe the twain shall meet by virtue of each moving in the direction of the other. East and West may contribute to a blended world where social and cognitive aspects of both regions are represented but transformed—like the individual ingredients in a stew that are recognizable but are altered as they alter the whole. It may not be too much to hope that this stew will contain the best of each culture.

NOTES

Introduction

xiv   *A dozen years before:* Nisbett and Ross (1980).
xv   *On the other hand:* Nisbett (1992); Nisbett, Fong, Lehman, and Cheng (1987).

Chapter 1: The Syllogism and the Tao

2   *The Greeks, more than:* Cromer (1993); Hamilton (1930/1973).
2   *One definition of happiness:* Hamilton (1930/1973), p. 25.
3   *The Greek sense of agency:* Galtung (1981).
4   *St. Luke said of the Athenians:* Hamilton (1930/1973), p. 33.
4   *Whereas many great contemporary:* Lin (1936); Toulmin and Goodfield (1961), p. 84.
5   *. . . for the early Confucians:* Rosemont (1991), p. 90.
5   *The ideal of happiness:* Lin (1936), p.121.
6   *Individual rights in China:* Munro (1985).
6   *As the British philosopher:* Lloyd (1990), p. 550.
7   *But when they discovered:* Nakayama (1969).
7   *The Chinese have been credited:* Logan (1986), p. 51.
8   *But, as philosopher Hajime Nakamura:* Nakamura (1964/1985).
8   *And as philosopher:* Munro (1969), p. 55.
9   *If the senses seemed:* Lloyd (1990), pp. 117–18.
11   *As communications theorist Robert Logan:* Logan (1986).
15   *It was the religion:* Lin (1936), p. 117.
15   *The ubiquitous word was ch'i:* Lin (1936), p. 122.
16   *Confucian-inspired themes:* Lin (1936), pp. 119–20.
16   *There is an adage:* Lin (1936), p. 117.
17   *They will share this mind:* Munro (1985), p. 119.

17 *If the emperor does:* Munro (1985).
18 *In Chinese literary criticism:* Lin (1936), p. 83.
18 *For the Chinese, the background scheme:* Hansen (1983), p. 31.
18 *"Their universe was a":* Needham (1962), p. 14.
19 *The world was complicated:* Munro (1985), p. 19.
20 *The greatest of all:* Lloyd (1991).
20 *A plausible account of:* Cromer (1993), p. viii.
21 *As philosopher Yu-lan Fung:* Fung (1983).
22 *They had knowledge of magnetism:* Needham (1962), p. 60.
22 *Health was dependent on:* Hadingham (1994).
24 *Thus any attempt to categorize:* Logan (1986), p. 122; Moser (1996), p. 116.
24 *The Chinese were right:* Nakamura (1964/1985), p. 189.
24 *Only the Greeks made:* Atran (1998).
25 *Whether this story is apocryphal:* Needham (1962).
26 *Like the ancient Chinese:* Lin (1936), p. 90.
26 *Chinese philosopher Mo-tzu:* Chan (1967), p. 47.
26 *Except for that brief interlude:* Becker (1986), p. 83.
26 *India did have a strong:* Becker (1986), p. 84.
26 *Algebra did not become:* Cromer (1993), p. 89.
27 *On the contrary:* Chang, cited in Becker (1986); Mao (1937/1962).

Chapter 2: The Social Origins of Mind

32 *It is essentially a distillation:* Barry, Child, and Bacon (1959); Berry (1976); Cole, Gay, Glick, and Sharp (1971); Cole and Scribner (1974); Cromer (1993); Nakamura (1964/1985); Needham (1954); Vygotsky (1930/1971), (1978); Whiting and Whiting (1975); Witkin and Berry (1975).
35 *The soil and climate:* McNeil (1962).
35 *As social psychologists Hazel Markus:* Markus and Kitayama (1991b), p. 246.
37 *"Science, in this view:"* Cromer (1993), p. 144.
40 *Even modern Chinese philosophers:* Lloyd (1990), pp. 124, 130; Tweed and Lehman (2002).
42 *Herman Witkin and his colleagues:* Witkin, Dyk, Faterson, Goodenough, and Karp (1974).
43 *And in fact this:* Berry and Annis (1974); Witkin and Berry (1975).
43 *And this is also:* Witkin and Berry (1975).
43 *To test this hypothesis:* Dershowitz (1971).
44 *And in fact relatively:* Witkin (1969).
44 *Field dependent people also:* Witkin and Goodenough (1977).
44 *and for social words:* Eagle, Goldberger, and Breitman (1969).
44 *And, when given their choice:* Greene (1973).
44 *But the implications of:* see Nisbett, Peng, Choi, and Norenzayan (2001) for a more formal treatment of the theory and its predictions.

Chapter 3: Living Together vs. Going It Alone

48 *The social-psychological characteristics:* For reviews of the social differences between East Asians and Westerners, see Fiske, Kitayama, Markus, and Nisbett (1998); Hsu (1981); Markus and Kitayama (1991b); Triandis (1995).

50 *As philosopher Hu Shih:* Shih (1919), p. 116, cited in King (1991).

50 *Anthropologist Edward T. Hall:* Hall (1976).

50 *As philosopher Donald Munro:* Munro (1985).

51 *I have presented a schematic:* Iyengar, Lepper, and Ross (1999).

53 *Such practices reflect:* The generic "I" is commonly used in China today, but this is a recent development following the Sun Yat-sen revolution in the early twentieth century.

53 *North Americans will tell:* Holmberg, Markus, Herzog, and Franks (1997).

53 *A study asking Japanese:* Cousins (1989).

53 *When describing themselves, Asians:* Ip and Bond (1995).

53 *Another study found:* Markus and Kitayama (1991b).

54 *On question after question:* Markus and Kitayama (1991b).

54 *Social psychologists Heejung Kim:* Kim and Markus (1999).

54 *Americans are much more likely:* Holmberg, et al. (1997).

54 *Asians rate themselves:* e.g., Kitayama, Markus, and Lieberman (1995); Bond and Cheung (1983).

55 *Contrast this with the American:* For a recent review of self-esteem East and West, see Heine, Lehman, Markus, and Kitayama (1999).

55 *An experiment by Steven Heine:* Heine, et al. (2001).

56 *The distinction is similar:* Tönnies (1887/1988).

57 *The Gemeinschaft is often:* Hofstede (1980); Hsu (1953; 1981); Triandis (1972; 1995).

57 *The terms "interdependent" and "independent":* Markus and Kitayama (1991b).

57 *Whereas it is common:* Shweder, Balle-Jensen, and Goldstein (in press).

58 *Social psychologists Sheena Iyengar:* Iyengar and Lepper (1999).

59 *Japanese mothers are particularly likely:* Azuma (1994); Fernald and Morikawa (1993).

59 *For example, Jeffrey Sanchez-Burks:* Sanchez-Burks, et al. (2002).

60 *The Japanese students reported:* Masuda and Nisbett (2001).

60 *Similarly, Kaiping Peng and Phoebe Ellsworth:* Unpublished study by Kaiping Peng and Phoebe Ellsworth, 2002.

61 *At first the child:* The example is owing to H. Kojima (1984).

61 *There are many ways:* For good theoretical treatments, see in particular Doi (1971/1981); Hampden-Turner and Trompenaars (1993); Hofstede (1980); Hsu (1953); Markus and Kitayama (1991a); and Triandis (1994a; 1995).

62 *He found dramatic:* Hofstede (1980).

62 *Similar data have been:* Hampden-Turner and Trompenaars (1993).

66 *But to the Australians:* Hampden-Turner and Trompenaars (1993), p. 123.

66 *Marketing experts Sang-pil Han:* Han and Shavitt (1994).
67 *Psychologists Wendi Gardner:* Gardner, Gabriel, and Lee (1999).
67 *In an "interdependent" version:* The method was developed by Trafimow and his colleagues(1991).
68 *Other evidence shows that:* Heine and Lehman (1997).
68 *Japanese who live in:* Kitayama, Markus, Matsumoto, and Norasakkunit (1997).
70 *Moreover, these differences among:* Sowell (1978).
72 *The sociologist Robert Bellah:* Bellah (1957/1985); Dien (1997; 1999); Lin (1936); Nakamura (1964/1985).
72 *Dora Dien has written:* Dien (1999), p. 377.
72 *This is amae:* Doi (1971/1981; 1974).
75 *More typically, the disputants:* Leung (1987).
75 *Negotiation also has:* See Cohen (1997) for a review of the differences between Western and Eastern negotiating styles.
75 *This view implies:* Kinhide (1976), pp. 45–46, cited in Cohen (1997).
76 *The Japanese awase:* Kinhide (1976), p. 40, cited in Cohen (1997).
76 *There is a belief:* Cohen (1997), p. 37.
77 *As nearly as we can tell:* Which is not to imply that marked differences have been present continuously. For example, no one would say that the eleventh-century European peasant was much of an individualist and both China and Japan have passed through periods in which individualism was highly valued, at least for artists and intellectuals.

Chapter 4: "Eyes in Back of Your Head" or "Keep Your Eye on the Ball"?

81 *Cognitive psychologists Mutsumi Imae:* Imae and Gentner (1994).
83 *Beginning in the late eighteenth:* Bradd Shore (1996) has an intriguing account of modularity in the West.
83 *In their survey:* Hampden-Turner and Trompenaars (1993).
86 *Li-jun Ji, Norbert Schwarz:* Ji, Schwarz, and Nisbett (2000).
87 *A research team from our labs:* Hedden, et al. (2000).
87 *There was no difference:* The difference was statistically significant for the young people only. The elderly showed a strong but not significant trend in the same direction as did the younger people.
87 *Developmental psychologists Han:* Han, Leichtman, and Wang (1998).
88 *They asked North American students:* Cohen and Gunz (2002).
89 *He achieved this by using:* Masuda and Nisbett (2001).
90 *The ability of the Japanese:* The concept of stimulus binding in perception is owing to Chalfonte and Johnson (1996).
92 *Short of the two men:* Simons and Levin (1997).
93 *In order to examine:* Masuda and Nisbett (2002).
95 *Exploring this question:* Ji, Peng, and Nisbett (2000).
95 *For other trials:* The moderate association corresponded to a correlation of .40; the strong association to a correlation of .60.
95 *Most strikingly, Americans:* Yates and Curley (1996).
96 *Ji, Peng, and I:* Ji, Peng, and Nisbett (2000).

96   *We presented East Asians:* Witkin, et al. (1954).
97   *Surveys show that Asians:* Sastry and Ross (1998).
97   *Social psychologists Beth Morling:* Morling, Kitayama, and Miyamoto (in press).
97   *A survey of Asians:* Sastry and Ross (1998).
98   *The managers thought:* Earley (1989).
98   *The adage that:* Yamaguchi, Gelfand, Mizuno, and Zemba (1997).
100 *Ellen Langer, a social psychologist:* Langer (1975).
100 *The illusion can sometimes:* Glass and Singer (1973).
101 *Ji, Peng, and I:* Ji, Peng, and Nisbett (2000).
104 *With Li-jun Ji:* Ji, Su, and Nisbett (2001).

Chapter 5: "The Bad Seed" or "The Other Boys Made Him Do It"?

112 *In order to be sure:* Morris and Peng (1994).
114 *The first cross-cultural study:* Miller (1984).
115 *Organizational psychologist Fiona Lee:* Lee, Hallahan, and Herzog (1996).
116 *The attributions of Hong Kong athletes:* Quotations provided in personal communication by Fiona Lee.
116 *Morris and Peng showed:* Morris and Peng (1994).
116 *They showed abstract cartoons:* Peng and Knowles (in press); Peng and Nisbett (2000).
118 *Ying-yi Hong and her colleagues:* Hong, Chiu, and Kung (1997).
119 *Peng and his colleague:* Peng and Knowles (in press).
119 *Ara Norenzayan, Incheol Choi:* Norenzayan, Choi, and Nisbett (2002).
120 *For example, we asked:* Erdley and Dweck (1993).
120 *Social psychologists Michael Morris:* Morris, et al. (1999).
122 *These same factors tend:* Leung, Cheung, Zhang, Song, and Dong (in press); McRae, Costa, and Yik (1996); Piedmont and Chae (1997); Yang and Bond (1990).
122 *Cultural psychologists Kuo-shu Yang:* Yang and Bond (1990).
122 *In a subsequent effort:* Cheung, et al. (in press); Cheung, Leung, Law, and Zhang (1996).
123 *As a consequence:* Ross (1977). Sometimes the FAE is called "the correspondence bias," meaning that people infer traits or attitudes corresponding to behavior (Gilbert and Malone, 1995). This term tends to be used when it can't be proved that the dispositional inference in question is a literal error, as opposed to just a preference for a particular type of explanation.
124 *If so, both you:* This experiment has actually been conducted. Students offered a lot of money for showing people around campus are likely to do it; students offered only a small amount of money are much less likely to do so. But observers of the behavior assume in the first case that they are watching a person who is generous with her time and in the second case that they are watching a person who is very disinclined to lend a helping hand. Nisbett, Caputo, Legant, and Maracek (1973).

124   *The first solid experimental:* For example, Jones and Harris (1967).
125   *Chinese, Japanese, and Koreans:* Choi and Nisbett (1998); Kitayama and Masuda (1997); Krull, et al. (1996).
125   *Incheol Choi and I:* Choi and Nisbett (1998).
125   *Ara Norenzayan, Incheol Choi:* Norenzayan, et al. (2002).
127   *Historian Masako Watanabe:* Watanabe (1998).
128   *Consistent with the lesser complexity:* Choi, Dalal, and Kim-Prieto (2000).
130   *Cognitive psychologist Baruch Fischhoff:* Fischhoff (1975).
131   *Incheol Choi and I:* Choi (1998); Choi and Nisbett (2000).
133   *The fine young men:* Darley and Batson (1973).

Chapter 6: Is the World Made Up of Nouns or Verbs?

137   *Jorge Luis Borges, the Argentine writer:* Borges (1966).
138   *Change in* wind *would:* Munro (1969), p. 41.
138   *They were simply not concerned:* Moser (1996), p. 171.
138   *For the ancient Taoist philosopher:* Mote (1971), p. 102.
138   *The five colors cause:* cited in Hansen (1983), p. 108.
139   *Finding the features:* Chan (1967); Hansen (1983), p. 34; Moser (1996), p. 171.
140   *Chiu found that the American:* Chiu (1972).
140   *Li-jun Ji, Zhiyong Zhang, and I:* Ji, Nisbett, and Zhang (2002).
142   *To test this possibility:* Norenzayan (1999); Norenzayan, Smith, Kim, and Nisbett (in press).
144   *In order to test:* Norenzayan, et al. (in press).
144   *We told participants that:* This experiment is based on procedures developed by Allen and Brooks (1991).
146   *Most Westerners who have:* Osherson, Smith, Wilkie, Lopez, and Shafir (1990).
147   *Incheol Choi, Edward E. Smith, and I:* Choi, Nisbett, and Smith (1997).
149   *"Verbs," says cognitive psychologist:* Gentner (1981), p. 168.
149   *Given these differences:* Gentner (1982).
149   *Developmental psycholinguist Twila Tardif:* Tardif (1996).
149   *First, verbs are more salient:* Gopnick and Choi (1990); Tardif (1996).
150   *Developmental psychologists Anne Fernald:* Fernald and Morikawa (1993).
150   *An American mother's patter:* Fernald and Morikawa (1993), p. 653.
151   *Developmental psychologists Linda Smith:* Smith, Jones, Landau, Gershkoff-Stowe, and Samuelson (2002).
151   *Developmental psychologists Susan Gelman:* Gelman and Tardif (1998).
152   *They found that object-naming:* Gopnick and Choi (1990).
153   *In the East:* Stevenson and Lee (1996).
154   *Ludwig Wittgenstein, in his* Philosophical Investigations: Skepticism about necessary and sufficient conditions was present, however, as early as the Scottish Enlightenment.
156   *"Generic" noun phrases:* Lucy (1992).

156 *The philosopher David Moser:* Moser (1996).
157 *The linguistic anthropologist:* Heath (1982).
158 *It is difficult for Japanese:* Cousins (1989).
159 *To English speakers:* Twila Tardif pointed out this amusing language difference, arbitrary from an information-processing standpoint, but essential from a linguistic standpoint.
159 *According to linguistic anthropologists:* Whorf (1956).
159 *Recall that Li-jun Ji:* Ji, Zhang, and Nisbett (2002).
159 *Psycholinguists make a distinction:* Ervin and Osgood (1954); Lambert, Havelka, and Crosby (1958).

Chapter 7: "Ce N'est Pas Logique" or "You've Got a Point There"?

165 *" . . . The most striking difference":* Graham (1989), p. 6.
165 *" . . . It is precisely because":* Liu (1974).
165 *"The aim of the Chinese":* Lin (1936), p. 109.
166 *"To argue with logical consistency":* Nagashima (1973), p. 96.
166 *The Logicians in fact:* Chan (1967).
166 *The Mo-tzu tradition:* Disheng (1990–91), p. 49.
166 *Nevertheless, even the Mohists:* Disheng (1990–91), p. 51; Lloyd (1990), p. 119.
166 *Moreover, despite Mohists' advances:* Disheng (1990–91), p. 51.
167 *So it seems to be:* Disheng (1990–91), p. 52.
167 *Mo-tzu wanted:* Chan (1967a); Disheng (1990–91), p. 51.
168 *Two studies by Ara Norenzayan:* Norenzayan (1999); Norenzayan, Smith, Kim, and Nisbett (in press).
168 *Prior experience makes them:* Sloman (1996).
169 *We asked Korean, Asian American:* Norenzayan, et al. (in press).
170 *We presented Korean and American:* Norenzayan, et al. (in press).
170 *The difference between:* It should be noted that we also looked at participants' judgments about the validity of *invalid* arguments. Koreans and Americans were equally influenced by conclusion plausibility for these arguments. I have no idea why.
171 *William McGuire showed that:* McGuire (1967).
171 *Ara Norenzayan and Beom Jun Kim:* Norenzayan and Kim (2002).
173 *Kaiping Peng and I:* Peng (1997); Peng and Nisbett (1999).
174 *At the risk of doing violence:* Peng (1997).
174 *Because reality is in constant flux:* Cao (1982); Liu (1988); Wang (1979).
175 *In fact opposites complete:* Chan (1967), p. 54.
175 *As the founder of the Taoist:* Lao-Zi (1993), p. 16.
175 *Or as Mao Tse-tung:* Mao (1937/1962), p. 42.
176 *There is a strong presumption:* Lin (1936), p. 110.
176 *Though the Hegelian:* Peng and Knowles (in press).
178 *In another study, Peng and I:* Peng and Nisbett (1999).
181 *In order to examine this question:* Peng and Nisbett (1999).
183 *There is evidence in fact:* Yates, Lee, and Bush (1997).

183 *Organizational psychologists Briley:* Briley, Morris, and Simonson (2000).
185 *This tendency mirrors:* Korzybyski (1933/1994); Lin (1936); Liu (1974); Nagashima (1973); Saul (1992).
185 *Incheol Choi argued that:* Choi (2001).
187 *Kaiping Peng and his colleagues:* Peng, Keltner, and Morikawa (2002).
188 *Organizational psychologists Richard Bagozzi:* Bagozzi, Wong, and Yi (1999).
189 *In research with young:* Geary, Salthouse, Chen, and Fan (1996); Hedden, et al. (in press).
189 *Teacher training in the East:* Stevenson and Stigler (1992).
189 *Both in America and in Asia:* Stevenson and Lee (1996).

Chapter 8: And If the Nature of Thought Is Not Everywhere the Same?

194 *Chinese justice is an art:* Lin (1936), p. 80.
194 *Japanese managers tend to:* Ohbuchi and Takahashi (1994).
195 *In the decade of the nineties:* French (2001). It should be noted, though, that a good many of the American Nobelists were born in some other country.
195 *That kind of thing happens:* French (2001).
197 *Recall the bitterness:* Hampden-Turner and Trompenaars (1993).
197 *But looked at as strengthening:* Hampden-Turner and Trompenaars (1993), pp. 123–24.
197 *Political scientist Peter Hays Gries:* Gries and Peng (2001).
198 *For the Chinese, any conception:* Munro (1985).
199 *One can be a Confucian:* Chan (1967), p. 31.
200 *On the other hand:* Dyson (1998).
202 *The philosopher Nelson Goodman:* Goodman (1965). The term is philosopher John Rawls's, but the concept is Goodman's.
202 *Philosopher Stephen Stich:* Stich (1990).
203 *As the philosopher S. H. Liu:* Liu (1974), p. 325.
203 *Many such schemas:* Basseches (1980); Riegel (1973).
204 *Some examples from Basseches's:* Basseches (1984).
205 *Two-valued Logic:* The concept of "two-valued logic" is owing to the General Semantics movement initiated by Alfred Korzybyski (1933–1994) and popularized in the United States by Berkeley professor S. I. Hayakawa (later a conservative U.S. Senator from California). During the 1950s and 1960s young Westerners with intellectual aspirations were given to wearing buttons emblazoned with "Null-A"— for non-Aristotelian thinking. It is probably no accident that an Eastern European and an Asian American were among the leaders of this antiformalistic logic movement. Though I find much of value in the stance they represented, they claimed far too much: Wars and insanity could be made a thing of the past if only people would realize that propositions need not be either true nor false.
206 *Developmental psychologists Joan Miller:* Miller and Bersoff (1995).

207 *The work is also important:* Doris (2002); Harman (1998–1999); Vranas (2001).
208 *We have seen that Easterners:* Choi (2001).
209 *Statistician and psychologist Robert Abelson:* Abelson (1995).
210 *Americans are so used to hearing:* Watanabe (1998).
210 *This view is shared:* Wilgoren (2001).
210 *A few years ago, Heejung Kim:* Kim (in press).
212 *Leaving aside for the moment:* Meyer and Kieras (1997).
213 *With a research team headed:* Park, et al. (2002).
216 *Just such an assertion:* Herrnstein and Murray (1994).

Epilogue

219 *Fukuyama has written:* Fukuyama (1992).
219 *Huntington has pronounced the world:* Huntington (1996).
220 *"In the emerging world":* Huntington, cited in Kaplan (2001).
221 *Harold Stevenson and his colleagues:* Personal communication from Harold Stevenson.
221 *A few years ago Kaiping Peng:* Peng, Nisbett, and Wong (1997).
222 *The odd results:* Heine, Lehman, Peng, and Greenholtz (2002).
224 *As Huntington has observed:* Huntington (1996).
226 *Recall the studies on causal perception:* Hong, Chiu, and Kung (1997).
227 *Similarly, Asian Americans answered:* Peng and Knowles (in press).
227 *Shinobu Kitayama and his colleagues:* Kitayama, Duffy, and Kawamura (2003).
228 *For example, they tried:* Kühnen, Hannover, and Schubert (2000).
228 *Kühnen and Daphna Oyserman:* Kühnen (2002).

REFERENCES

Abelson, R. P. (1995). *Statistics as Principled Argument*. Hillsdale, NJ: Lawrence Erlbaum.

Allen, S. W., and Brooks, L. R. (1991). Specializing in the operation of an explicit rule. *Journal of Experimental Social Psychology, General 120*, 3–19.

Atran, S. (1998). "Folk biology and the anthropology of science: Cognitive universals and cultural particulars." *Behavioral and Brain Sciences 21*, 547–569.

Azuma, H. (1994). *Education and Socialization in Japan*. Tokyo: University of Tokyo Press.

Bagozzi, R. P., Wong, N., and Yi, Y. (1999). "The role of culture and gender in the relationship between positive and negative affect." *Cognition and Emotion 13*, 641–672.

Barry, H., Child, I., and Bacon, M. (1959). Relation of child training to subsistence economy. *American Anthropologist 61*, 51–63.

Basseches, M. (1980). "Dialectical schemata: A framework for the empirical study of the development of dialectical thinking." *Human Development 23*, 400–421.

———. (1984). *Dialectical Thinking and Adult Development*. New Jersey: Ablex.

Becker, C. B. (1986). "Reasons for the lack of argumentation and debate in the Far East." *International Journal of Intercultural Relations 10*, 75–92.

Bellah, R. (1957/1985). *Tokagawa Religion: The Cultural Roots of Modern Japan*. New York: Free Press.

Berry, J. W. (1976). *Human Ecology and Cognitive Style: Comparative Studies in Cultural and Psychological Adaptation*. New York: Sage/Halsted.

Berry, J. W., and Annis, R. C. (1974). "Ecology, culture and differentiation." *International Journal of Psychology 9*, 173–193.

Bond, M. H., and Cheung, T. S. (1983). "College students' spontaneous self-concept: The effect of culture among respondents in Hong Kong, Japan, and the United States." *Journal of Cross-Cultural Psychology 14*, 153–171.

Borges, J. L. (1966). *Other Inquisitions 1937–1952*. New York: Washington Square Press.

Briley, D. A., Morris, M., and Simonson, I. (2000). "Reasons as carriers of culture: Dynamic vs. dispositional models of cultural influence on decision making." *Journal of Consumer Research 27*, 157–178.

Cao, C. J. (1982). *Explanation of Zhung Zi*. Beijing: Zhong Hua Publishing House.

Chalfonte, B. L., and Johnson, M. K. (1996). "Feature memory and binding in young and older adults." *Memory and Cognition 24*, 403–416.

Chan, W. T. (1967). "The story of Chinese philosophy." In C. A. Moore (ed.), *The Chinese Mind: Essentials of Chinese Philosophy and Culture*. Honolulu: East-West Center Press.

———. (1967). "Chinese theory and practice, with special reference to humanism." In C. A. Moore (ed.), *The Chinese Mind: Essentials of Chinese Philosophy and Culture*. Honolulu: East-West Center Press.

Cheung, F. M., Leung, K., Fang, R. M., Song, W. Z., Zhang, J. X., and Zhang, J. P. (in press). "Development of the Chinese personality assessment inventory." *Journal of Cross-Cultural Psychology*.

Cheung, F. M., Leung, K., Law, J. S., and Zhang, J. X. (1996). "Indigenous Chinese Personality Constructs." Paper presented at the XXVI International Congress of Psychology, Montreal, Canada.

Chiu, L.-H. (1972). "A cross-cultural comparison of cognitive styles in Chinese and American children." *International Journal of Psychology 7*, 235–242.

Choi, I. (1998). The cultural psychology of surprise: Holistic theories, contradiction, and epistemic curiosity. Unpublished Ph.D. thesis, University of Michigan, Ann Arbor.

———. (2001). The conflicted culture or who reads fortune-telling? Unpublished manuscript, Seoul National University.

Choi, I., Dalal, R., and Kim-Prieto, C. (2000). Information search in causal attribution: Analytic vs. holistic. Unpublished manuscript, Seoul National University.

Choi, I., and Nisbett, R. E. (1998). "Situational salience and cultural differences in the correspondence bias and in the actor-observer bias." *Personality and Social Psychology Bulletin 24*, 949–960.

———. (2000). "The cultural psychology of surprise: Holistic theories and recognition of contradiction." *Journal of Personality and Social Psychology 79*, 890–905.

Choi, I., Nisbett, R. E., and Smith, E. E. (1997). "Culture, categorization and inductive reasoning." *Cognition 65*, 15–32.

Cohen, D., and Gunz, A. (2002). As seen by the other . . . : The self from the "outside in" and the "inside out" in the memories and emotional perceptions of Easterners and Westerners. Unpublished manuscript: University of Illinois.

Cohen, R. (1997). *Negotiating Across Cultures: International Communication in an Interdependent World*. Washington, D.C.: United States Institute of Peace Press.

Cole, M., Gay, J., Glick, J. A., and Sharp, D. W. (1971). *The Cultural Context of Learning and Thinking*. New York: Basic Books.

Cole, M., and Scribner, S. (1974). *Culture and Thought: A Psychological Introduction*. New York: Wiley.

Cousins, S. D. (1989). "Culture and self-perception in Japan and the United States." *Journal of Personality and Social Psychology 56*, 124–131.

Cromer, A. (1993). *Uncommon Sense: The Heretical Nature of Science.* New York: Oxford University Press.

Darley, J. M., and Batson, C. D. (1973). "From Jerusalem to Jericho: A study of situational and dispositional variables in helping behavior." *Journal of Personality and Social Psychology 27,* 100–119.

Dershowitz, Z. (1971). "Jewish subcultural patterns and psychological differentiation." *International Journal of Psychology 6,* 223–231.

Diamond, J. (1997). *Guns, Germs, and Steel: The Fates of Human Societies.* New York: Norton.

Dien, D. S.-f. (1997). *Confucianism and Cultural Psychology: Comparing the Chinese and the Japanese.* Hayward, CA: California State University.

———. (1999). "Chinese authority-directed orientation and Japanese peer-group orientation: Questioning the notion of collectivism." *Review of General Psychology 3,* 372–385.

Disheng, Y. (1990–91). "China's traditional mode of thought and science: A critique of the theory that China's traditional thought was primitive thought." *Chinese Studies in Philosophy,* Winter, 43–62.

Doi, L. T. (1971/1981). *The Anatomy of Dependence* (2nd ed.). Tokyo: Kodansha.

———. (1974). "*Amae:* A key concept for understanding Japanese personality structure." In R. J. Smith and R. K. Beardsley (eds.), *Japanese Culture: Its Development and Characteristics.* Chicago: Aldine.

Doris, J. M. (2002). *Lack of Character: Personality and Moral Behavior.* New York: Cambridge University Press.

Dyson, F. J. (1998, May 28). "Is God in the Lab?" *New York Review of Books,* pp. 8–10.

Eagle, M., Goldberger, L., and Breitman, M. (1969). "Field dependence and memory for social vs. neutral and relevant vs. irrelevant incidental stimuli." *Perceptual and Motor Skills 29,* 903–910.

Earley, P. C. (1989). "East meets west meets mideast: Further explorations of collectivistic and individualistic work groups." *Academy of Management Journal 36,* 565–581.

Erdley, C. A., and Dweck, C. S. (1993). "Children's implicit personality theories as predictors of their social judgments." *Child Development 64,* 863–878.

Ervin, S. M., and Osgood, C. E. (1954). "Second language learning and bilingualism." *Journal of Abnormal and Social Psychology 49,* Supplement, 139–146.

Fernald, A., and Morikawa, H. (1993). "Common themes and cultural variations in Japanese and American mothers' speech to infants." *Child Development 64,* 637–656.

Fischhoff, B. (1975). "Hindsight ≠ Foresight: The effect of outcome knowledge on judgment under uncertainty." *Journal of Experimental Psychology: Human Perception and Performance 1,* 288–299.

Fiske, A. P., Kitayama, S., Markus, H. R., and Nisbett, R. E. (1998). "The cultural matrix of social psychology." In D. T. Gilbert, S. T. Fiske, and G. Lindzey (eds.), *Handbook of Social Psychology* (4th ed.), pp. 915–981. Boston: McGraw-Hill.

French, H. W. (2000, May 2). "Japan debates culture of covering up." *New York Times,* p. A12.

Fukuyama, F. (1992). *The End of History and the Last Man.* New York: Free Press.

Fung, Y. (1983). *A History of Chinese Philosophy* (D. Bodde, trans., vol. 1–2). Princeton: Princeton University Press.

Galtung, J. (1981). "Structure, culture, and intellectual style: An essay comparing saxonic, teutonic, gallic and nipponic approaches." *Social Science Information 20*, 817–856.

Gardner, W. L., Gabriel, S., and Lee, A. Y. (1999). " 'I' value freedom, but 'we' value relationships: Self-construal priming mirrors cultural differences in judgment." *Psychological Science 10*, 321–326.

Geary, D. C., Salthouse, T. A., Chen, G-P., and Fan, L. (1996). "Are East Asian versus American differences in arithmetical ability a recent phenomenon?" *Developmental Psychology 32*, 254–262.

Gelman, S. A., and Tardif, T. (1998). "A cross-linguistic comparison of generic noun phrases in English and Mandarin." *Cognition 66*, 215–248.

Gentner, D. (1981). "Some interesting differences between nouns and verbs." *Cognition and Brain Theory 4*, 161–178.

———. (1982). "Why nouns are learned before verbs: Linguistic relativity vs. natural partitioning." In S. A. Kuczaj, ed., *Language Development: Vol. 2. Language, Thought and Culture.* Hillsdale, NJ: Lawrence Erlbaum.

Gilbert, D. T., and Malone, P. S. (1995). "The correspondence bias." *Psychological Bulletin 117*, 21–38.

Glass, D. C., and Singer, J. E. (1973). "Experimental studies of uncontrollable and unpredictable noise." *Representative Research in Psychology 4*, 165–183.

Goodman, N. (1965). *Fact, Fiction and Forecast* (2nd ed.). Indianapolis: Bobbs-Merrill.

Gopnik, A., and Choi, S. (1990). "Do linguistic differences lead to cognitive differences? A cross-linguistic study of semantic and cognitive development." *First Language 10*, 199–215.

Graham, A. C. (1989). *Disputers of the Tao.* La Salle: Open Court Press.

Greene, L. R. (1973). "Effects of field independence, physical proximity and evaluative feedback, affective reactions and compliance in a dyadic interaction." *Dissertation Abstracts International 34*, 2284–2285.

Gries, P. H., and Peng, K. (2002). "Culture clash? Apologies East and West." *Journal of Contemporary China 11*, 173–178.

Hadingham, E. (1994). "The mummies of Xinjiang." *Discover 15*, 68–77.

Hall, E. T. (1976). *Beyond Culture.* New York: Anchor Books.

Hamilton, E. (1930/1973). *The Greek Way.* New York: Avon.

Hampden-Turner, C., and Trompenaars, A. (1993). *The Seven Cultures of Capitalism: Value Systems for Creating Wealth in the United States, Japan, Germany, France, Britain, Sweden, and the Netherlands.* New York: Doubleday.

Han, J. J., Leichtman, M. D., and Wang, Q. (1998). "Autobiographical memory in Korean, Chinese, and American children." *Developmental Psychology 34*, 701–713.

Han, S., and Shavitt, S. (1994). "Persuasion and culture: Advertising appeals in individualistic and collectivistic societies." *Journal of Experimental Social Psychology 30*, 326–350.

Hansen, C. (1983). *Language and Logic in Ancient China.* Ann Arbor: University of Michigan Press.

Harman, G. (1998–1999). "Moral philosophy meets social psychology: Virtue ethics and the fundamental attribution error." *Proceedings of the Aristotelian Society 1998–99*, pp. 315–331.

Heath, S. B. (1982). "What no bedtime story means: Narrative skills at home and school." *Language in Society 11*, 49–79.

Hedden, T., Ji, L., Jing, Q., Jiao, S., Yao, C., Nisbett, R. E., and Park, D. C. (2000). Culture and age differences in recognition memory for social dimensions. Paper presented at the Cognitive Aging Conference, Atlanta.

Hedden, T., Park, D. C., Nisbett, R. E., Ji, L., Jing, Q., and Jiao, S. (2002). "Cultural variation in verbal versus spatial neuropsychological function across the lifespan." *Neuropsychology 16*, 65–73.

Heine, S. J., Kitayama, S., Lehman, D. R., Takata, T., Ide, E., Leung, C., and Matsumoto, H. (2001). "Divergent consequences of success and failure in Japan and North America: An investigation of self-improving motivation." *Journal of Personality and Social Psychology 81*, 599–615.

Heine, S. J., and Lehman, D. R. (1997). Acculturation and self-esteem change: Evidence for a Western cultural foundation in the construct of self-esteem. Paper presented at the second meeting of the Asian Association of Social Psychology, Kyoto, Japan.

Heine, S. J., Lehman, D. R., Markus, H. R., and Kitayama, S. (1999). "Is there a universal need for positive self-regard?" *Psychological Review 106*, 766–794.

Heine, S. J., Lehman, D. R., Peng, K., and Greenholtz, J. (2002). What's Wrong with Cross-cultural Comparisons of Subjective Likert Scales?: The Reference Group Effect. Unpublished manuscript, University of British Columbia, Vancouver, B.C. .

Herrnstein, R. J., and Murray, C. (1994). *The Bell Curve: Intelligence and Class Structure in American Life*. New York: The Free Press.

Hofstede, G. (1980). *Culture's Consequences: International Differences in Work-related Values*. Beverly Hills: Sage.

Holmberg, D., Markus, H., Herzog, A. R., and Franks, M. (1997). Self-making in American Adults: Content, Structure and Function. Unpublished manuscript, University of Michigan, Ann Arbor.

Hong, Y., Chiu, C., and Kung, T. (1997). "Bringing culture out in front: Effects of cultural meaning system activation on social cognition." In K. Leung, Y. Kashima, U. Kim, and S. Yamaguchi, eds., *Progress in Asian Social Psychology* 1. Singapore: Wiley, 135–146.

Hsu, F. L. K. (1953). *Americans and Chinese: Two Ways of Life*. New York: Schuman.

———. (1981). "The self in cross-cultural perspective." In A. J. Marsella, B. D. Vos, and F. L. K. Hsu, eds., *Culture and Self* (pp. 24–55). London: Tavistock.

Huntington, S. P. (1996). *The Clash of Civilizations and the Remaking of World Order*. New York: Simon & Schuster.

Imai, M., and Gentner, D. (1994). "A cross-linguistic study of early word meaning: Universal ontology and linguistic influence." *Cognition 62*, 169–200.

Ip, G. W. M., and Bond, M. H. (1995). "Culture, values, and the spontaneous self-concept." *Asian Journal of Psychology 1*, 29–35.

Iyengar, S. S., and Lepper, M. R. (1999). "Rethinking the role of choice: A cultural perspective on intrinsic motivation." *Journal of Personality and Social Psychology 76*, 349–366.

Iyengar, S. S., Lepper, M. R., and Ross, L. (1999). "Independence from whom? Interdependence from whom? Cultural perspectives on ingroups versus outgroups." In D. A. Prentice and D. T. Miller, eds., *Cultural Divides: Understanding and Overcoming Group Conflict.* New York: Russell Sage Foundation.

Ji, L., Peng, K., and Nisbett, R. E. (2000). "Culture, control, and perception of relationships in the environment." *Journal of Personality and Social Psychology 78*, 943–955.

Ji, L., Schwarz, N., and Nisbett, R. E. (2000). "Culture, autobiographical memory, and social comparison: Measurement issues in cross-cultural studies." *Personality and Social Psychology Bulletin 26*, 585–593.

Ji, L., Su, Y., and Nisbett, R. E. (2001). "Culture, prediction, and change." *Psychological Science 12*, 450–456.

Ji, L., Zhang, Z., and Nisbett, R. E. (2002). Culture, language and categorization. Unpublished manuscript, Queens University, Kingston, Ontario.

Jones, E. E., and Harris, V. A. (1967). "The attribution of attitudes." *Journal of Experimental Social Psychology 3*, 1–24.

Kaplan, R. D. (2001, December). "Looking the world in the eye." *Atlantic Monthly*, 68–82.

Kim, H. (in press). "We talk, therefore we think? A cultural analysis of the effect of talking on thinking." *Journal of Personality and Social Psychology.*

Kim, H., and Markus, H. R. (1999). "Deviance or uniqueness, harmony or conformity?: A cultural analysis." *Journal of Personality and Social Psychology 77*, 785–800.

King, A. Y.-c. (1991). "Kuan-hsi and network building: A sociological interpretation." *Daedelus 120*, 60–84.

Kinhide, M. (1976). "The cultural premises of Japanese diplomacy." In J. C. f. I. Exchange, ed., *The Silent Power: Japan's Identity and World Role.* Tokyo: Simul Press.

Kitayama, S., Duffy, S., and Kawamura, T. (2003). Perceiving an object in its context in different cultures: A cultural look at the New Look. Unpublished manuscript, Kyoto University, Kyoto.

Kitayama, S., Markus, H. R., and Lieberman, C. (1995). "The collective construction of self-esteem: Implications for culture, self, and emotion." In J. Russell, J. Fernandez-Dols, T. Manstead, and J. Wellenkamp, eds., *Everyday Conceptions of Emotion: An Introduction to the Psychology, Anthropology, and Linguistics of Emotion.* Dordrecht: Kluwer Academic Publishers.

Kitayama, S., Markus, H. R., Matsumoto, H., and Norasakkunit, V. (1997). "Individual and collective processes in the construction of the self: Self-enhancement in the United States and self-depreciation in Japan." *Journal of Personality and Social Psychology 72*, 1245–1267.

Kitayama, S., and Masuda, T. (1997). "Shaiaiteki ninshiki no bunkateki baikai model: taiousei bias no bunkashinrigakuteki kentou. (Cultural psychology of social inference: The correspondence bias in Japan.)." In K. Kashiwagi, S. Kitayama, and H. Azuma, eds., *Bunkashinrigaju: riron to jisho. (Cultural Psychology: Theory and Evidence).* Tokyo: University of Tokyo Press.

Kojima, H. (1984). "A significant stride toward the comparative study of control." *American Psychologist 39*, 972–973.

Korzybski, A. (1933/1994). *Science and Sanity: An Introduction to non-Aristotelian Systems and General Semantics.* Englewood, NJ: Institute of General Semantics.

Krull, D. S., Loy, M., Lin, J., Wang, C.-F., Chen, S., and Zhao, X. (1996). The fundamental attribution error: Correspondence bias in independent and interdependent cultures. Paper presented at the 13th Congress of the International Association for Cross-Cultural Psychology, Montreal, Quebec, Canada.

Kühnen, U., Hannover, B., Röder, U., Schubert, B., Shah, A. A., and Zakaria, S. (2000). "Cross-cultural variations in identifying embedded figures: Comparisons from the U.S., Germany, Russia and Malaysia." *Journal of Cross-Cultural Psychology 32*, 365–371.

Kühnen, U., Hannover, B., and Schubert, B. (2001). "The semantic-procedural interface model of the self: The role of self-knowledge for context-dependent versus context-independent modes of thinking." *Journal of Personality and Social Psychology 80*, 397–409.

Kühnen, U., and Oyserman, D. (2002). Thinking About the Self Influences Thinking in General: Cognitive Consequences of Salient Self-concept. Unpublished manuscript, University of Michigan, Ann Arbor.

Lambert, W. E., Havelka, J., and Crosby, C. (1958). "The influence of language acquisition contexts on bilingualism." *Journal of Abnormal and Social Psychology 56*, 239–244.

Langer, E. (1975). "The illusion of control." *Journal of Personality and Social Psychology 32*, 311–328.

Lao-Zi. (1993). *The Book of Lao Zi.* Beijing: Foreign Language Press.

Lee, F., Hallahan, M., and Herzog, T. (1996). "Explaining real life events: How culture and domain shape attributions." *Personality and Social Psychology Bulletin 22*, 732–741.

Leung, K. (1987). "Some determinants of reactions to procedural models for conflict resolution: A cross-national study." *Journal of Personality and Social Psychology 53*, 898–908.

Leung, K., Cheung, F. M., Zhang, J. X., Song, W. Z., and Dong, X. (in press). "The five factor model of personality in China." In K. Leung, Y. Kashima, U. Kim, and S. Yamaguchi, eds., *Progress in Asian Social Psychology* 1. Singapore: John Wiley.

Lin, Y. (1936). *My Country and My People.* London: William Heinemann.

Liu, S. H. (1974). "The use of analogy and symbolism in traditional Chinese philosophy." *Journal of Chinese Philosophy 1*, 313–338.

Liu, X. G. (1988). *The Philosophy of Zhung Zi and Its Evolution.* Beijing: The Social Science Press of China.

Lloyd, G. E. R. (1990). *Demystifying Mentalities.* New York: Cambridge University Press.

———. (1991). "The invention of nature." In G. E. R. Lloyd, ed., *Methods and Problems in Greek Science.* Cambridge: Cambridge University Press.

Logan, R. F. (1986). *The Alphabet Effect.* New York: Morrow.

Lucy, J. A. (1992). *Grammatical Categories and Cognition: A Case Study of the Linguistic Relativity Hypothesis.* New York: Cambridge University Press.

Mao, T.-t. (1937/1962). *Four Essays on Philosophy.* Beijing: People's Press.

Markus, H., and Kitayama, S. (1991a). "Cultural variation in the self-concept." In J. Strauss and G. R. Goethals, eds., *The Self: Interdisciplinary Approaches.* New York: Springer-Verlag.

――――. (1991b). "Culture and the self: Implications for cognition, emotion, and motivation." *Psychological Review 98*, 224–253.

Masuda, T., and Nisbett, R. E. (2001). "Attending holistically vs. analytically: Comparing the context sensitivity of Japanese and Americans." *Journal of Personality and Social Psychology 81*, 922–934.

――――. (2002). Change blindness in Japanese and Americans. Unpublished manuscript, University of Michigan, Ann Arbor.

McGuire, W. J. (1967). "Cognitive consistency and attitude change." In M. Fishbein, ed., *Attitude Theory and Measurement* (pp. 357–365). New York: John Wiley.

McNeil, W. H. (1962). *The Rise of the West: A History of the Human Community.* Chicago: University of Chicago Press.

McRae, R. R., Costa, P. T., and Yik, M. S. M. (1996). "Universal aspects of Chinese personality structure." In M. H. Bond, ed., *The Handbook of Chinese Psychology.* Hong Kong: Oxford University Press.

Meyer, D. E., and Kieras, D. E. (1997). "A computational theory of executive cognitive processes and multiple-task performance: I. Basic mechanisms." *Psychological Review 104*, 3–65.

Miller, J. G. (1984). "Culture and the development of everyday social explanation." *Journal of Personality and Social Psychology 46*, 961–978.

Miller, J. G., and Bersoff, D. M. (1995). "Development in the context of everyday family relationships: Culture, interpersonal morality, and adaptation." In M. Killen and D. Hart, eds., *Morality of Everyday Life: A Developmental Perspective* (pp. 259–282). Cambridge: Cambridge University Press.

Morling, B., Kitayama, S., and Miyamoto, Y. (in press). "Cultural practices emphasize influence in the U.S. and adjustment in Japan." *Personality and Social Psychology Bulletin.*

Morris, M., Leung, K., and Sethi, S. (1999). Person perception in the heat of conflict: Perceptions of opponents' traits and conflict resolution in two cultures. Unpublished manuscript, Stanford University.

Morris, M. W., and Peng, K. (1994). "Culture and cause: American and Chinese attributions for social and physical events." *Journal of Personality and Social Psychology 67*, 949–971.

Moser, D. J. (1996). Abstract thinking and thought in ancient Chinese and early Greek societies. Unpublished Ph.D. thesis, University of Michigan, Ann Arbor.

Mote, F. W. (1971). *Intellectual Foundations of China.* New York: Knopf.

Munro, D. (1985). Introduction. In D. Munro, ed., *Individualism and Holism: Studies in Confucian and Taoist Values* (pp. 1–34). Ann Arbor: Center for Chinese Studies, University of Michigan.

Munro, D. J. (1969). *The Concept of Man in Early China.* Stanford, CA: Stanford University Press.

Nagashima, N. (1973). "A reversed world: Or is it?" In R. Horton and R. Finnegan, eds., *Modes of Thought.* London: Faber and Faber.

Nakamura, H. (1964/1985). *Ways of Thinking of Eastern Peoples.* Honolulu: University of Hawaii Press.

Nakayama, S. (1969). *A History of Japanese Astronomy.* Cambridge, MA: Harvard University Press.

Needham, J. (1954). *Science and Civilisation in China,* Vol. 1. Cambridge, UK: Cambridge University Press.

———. (1962). *Science and Civilisation in China: Physics and Physical Technology,* Vol. 4. Cambridge, UK: Cambridge University Press.

Nisbett, R. E. (1992). *Rules for Reasoning.* Hillsdale, NJ: Lawrence Erlbaum.

Nisbett, R. E., Caputo, C., Legant, P., and Maracek, J. (1973). "Behavior as seen by the actor and as seen by the observer." *Journal of Personality and Social Psychology 27,* 154–164.

Nisbett, R. E., Fong, G. T., Lehman, D. R., and Cheng, P. W. (1987). "Teaching reasoning." *Science 238,* 625–631.

Nisbett, R. E., Peng, K., Choi, I., and Norenzayan, A. (2001). "Culture and systems of thought: Holistic vs. analytic cognition." *Psychological Review 108,* 291–310.

Nisbett, R. E., and Ross, L. (1980). *Human Inference: Strategies and Shortcomings of Social Judgment.* Englewood Cliffs, NJ: Prentice-Hall.

Norenzayan, A. (1999). Rule-based and experience-based thinking: The cognitive consequences of intellectual traditions. Unpublished Ph.D. thesis, University of Michigan, Ann Arbor, MI.

Norenzayan, A., Choi, I., and Nisbett, R. E. (2002). "Cultural similarities and differences in social inference: Evidence from behavioral predictions and lay theories of behavior." *Personality and Social Psychology Bulletin 28,* 109–120.

Norenzayan, A., and Kim, B. J. (2002). A cross-cultural comparison of regulatory focus and its effect on the logical consistency of beliefs. Unpublished manuscript, University of British Columbia, Vancouver, B.C.

Norenzayan, A., Smith, E. E., Kim, B. J., and Nisbett, R. E. (in press). "Cultural preferences for formal versus intuitive reasoning." *Cognitive Science.*

Ohbuchi, K. I., and Takahashi, Y. (1994). "Cultural styles of conflict management in Japanese and Americans: Passivity, covertness, and effectiveness of strategies." *Journal of Applied Psychology 24,* 1345–1366.

Osherson, D. N., Smith, E. E., Wilkie, O., Lopez, A., and Shafir, E. (1990). "Category-based induction." *Psychological Review 97,* 185–200.

Park, D., Hedden, T., Jing, Q., Shulan, J., Yao, C., and Nisbett, R. E. (2002). Culture and the aging mind. Unpublished manuscript, University of Michigan, Ann Arbor, MI.

Peng, K. (1997). Naive dialecticism and its effects on reasoning and judgment about contradiction. Unpublished Ph.D. thesis, University of Michigan, Ann Arbor, MI.

———. (2001). "Psychology of dialectical thinking." In N. J. Smelser and P. B. Baltes, eds., *International Encylopedia of the Social and Behavioral Sciences,* Vol. 6 (pp. 3634–3637). Oxford: Elsevier Science.

Peng, K., Keltner, D., and Morikawa, S. (2002). Culture and judgment of facial expression. Unpublished manuscript, University of California, Berkeley.

Peng, K., and Knowles, E. (in press). "Culture, ethnicity and the attribution of physical causality." *Personality and Social Psychology Bulletin.*

Peng, K., and Nisbett, R. E. (1999). "Culture, dialectics, and reasoning about contradiction." *American Psychologist 54,* 741–754.

Peng, K., and Nisbett, R. E. (2000). Cross-cultural Similarities and Differences in the Understanding of Physical Causality. Unpublished manuscript, University of California, Berkeley.

Peng, K., Nisbett, R. E., and Wong, N. (1997). "Validity problems of cross-cultural value comparison and possible solutions." *Psychological Methods 2,* 329–341.

Piedmont, R. L., and Chae, J. H. (1997). "Cross-cultural generalizability of the five-factor model of personality: Development and validation of the NEO-PI-R for Koreans." *Journal of Cross-Cultural Psychology 28,* 131–155.

Riegel, K. F. (1973). "Dialectical operations: The final period of cognitive development." *Human Development 18,* 430–443.

Rosemont, H., Jr. (1991). "Rights-bearing individuals and role-bearing persons." In M. I. Bockover, ed., *Rules, Rituals and Responsibility: Essays Dedicated to Herbert Fingarette.* LaSalle, IL: Open Court Press.

Ross, L. (1977). "The intuitive psychologist and his shortcomings." In L. Berkowitz, ed., *Advances in Experimental Social Psychology,* Vol. 10 (pp. 173–220). New York: Academic Press.

Sanchez-Burks, J., Lee, F., Choi, I., Nisbett, R. E., Zhao, S., and Koo, J. (2002). Conversing across cultural ideologies: East-West communication styles in work and non-work contexts. Unpublished manuscript, University of Southern California.

Sastry, J., and Ross, C. E. (1998). "Asian ethnicity and the sense of personal control." *Social Psychology Quarterly 61,* 101–120.

Saul, J. R. (1992). *Voltaire's Bastards: The Dictatorship of Reason in the West.* New York: Random House.

Shih, H. (1919). *Chung-kuo che-hsueh shi ta-kang (An Outline of the History of Chinese Philosophy).* Shanghai: Commercial Press.

Shore, B. (1996). *Culture in Mind: Cognition, Culture and the Problem of Meaning.* New York: Oxford University Press.

Shweder, R., Balle-Jensen, L., and Goldstein, W. (in press). "Who sleeps by whom revisited: A method for extracting the moral goods implicit in praxis." In P. J. Miller, J. J. Goodnow, and F. Kessell, eds., *Cultural Practices as Contexts for Development.* San Francisco: Jossey-Bass.

Simons, D. J., and Levin, D. T. (1997). "Change blindness." *Trends in Cognitive Sciences 1,* 261–267.

Sloman, S. (1996). "The empirical case for two systems of reasoning." *Psychological Bulletin 119,* 30–22.

Smith, L. B., Jones, S. S., Landau, B., Gershkoff-Stowe, L., and Samuelson, L. (2002). "Object name learning provides on-the-job training for attention." *Psychological Science 13,* 13–19.

Sowell, T., ed. (1978). *Essays and Data on American Ethnic Groups.* New York: The Urban Institute.

Stevenson, H. W., and Lee, S. (1996). "The academic achievement of Chinese students." In M. H. Bond, ed., *The Handbook of Chinese Psychology* (pp. 124–142). New York: Oxford University Press.

Stevenson, H. W., and Stigler, J. W. (1992). *The Learning Gap: Why Our Schools Are Failing and What We Can Learn from Japanese and Chinese Education.* New York: Summit Books.

Stich, S. (1990). *The Fragmentation of Reason.* Cambridge, MA: MIT Press.

Tardif, T. (1996). "Nouns are not always learned before verbs: Evidence from Mandarin-speakers early vocabularies." *Developmental Psychology* 32, 492–504.

Toulmin, S., and Goodfield, J. (1961). *The Fabric of the Heavens: The Development of Astronomy and Physics.* New York: Harper & Row.

Tönnies, F. (1887/1988). *Community and Society.* New Brunswick, Oxford: Transaction Books.

Trafimow, D., Triandis, H. C., and Goto, S. G. (1991). "Some tests of the distinction between the private self and the collective self." *Journal of Personality and Social Psychology* 60, 649–655.

Triandis, H. C. (1972). *The Analysis of Subjective Culture.* New York: Wiley.

———. (1994). *Culture and Social Behavior.* New York: McGraw-Hill.

———. (1995). *Individualism and Collectivism.* Boulder, CO: Westview Press.

Tweed, R. G., and Lehman, D. (2002). "Learning considered within a cultural context: Confucian and Socratic approaches." *American Psychologist* 57, 89–99.

Vranas, P. B. M. (2001). Respect for persons: An epistemic and pragmatic investigation. Unpublished Ph.D. thesis, University of Michigan, Ann Arbor, MI.

Vygotsky, L. S. (1930/1971). "The development of higher psychological functions." In J. Wertsch, ed., *Soviet Activity Theory.* Armonk, NY: Sharpe.

———. (1978). *Mind in Society: The Development of Higher Psychological Processes.* Cambridge: Harvard University Press.

Wang, D. J. (1979). *The History of Chinese Logical Thought.* Shanghai: People's Press of Shanghai.

Watanabe, M. (1998). Styles of reasoning in Japan and the United States: Logic of education in two cultures. Paper presented at the American Sociological Association, San Francisco, CA.

Weisz, J. R., Rothbaum, F. M., and Blackburn, T. C. (1984). "Standing out and standing in: The psychology of control in America and Japan." *American Psychologist* 39, 955–969.

Whiting, B. B., and Whiting, J. W. M. (1975). *Children of Six Cultures: A Psycho-cultural Analysis.* Cambridge: Harvard University Press.

Whorf, B. L. (1956). *Language, Thought and Reality.* New York: Wiley.

Wilgoren, J. (2001, August 9). "World of debating grows and Vermont is the lab." *New York Times*, p. A12.

Witkin, H. A.. (1969). *Social Influences in the Development of Cognitive Style.* New York: Rand McNally.

Witkin, H. A., and Berry, J. W. (1975). "Psychological differentiation in cross-cultural perspective." *Journal of Cross Cultural Psychology* 6, 4–87.

Witkin, H. A., Dyk, R. B., Faterson, H. F., Goodenough, D. R., and Karp, S. A. (1974). *Psychological Differentiation.* Potomac: Lawrence Erlbaum Associates.

Witkin, H. A., and Goodenough, D. R. (1977). "Field dependence and inter-personal behavior." *Psychological Bulletin 84,* 661–689.

Witkin, H. A., Lewis, H. B., Hertzman, M., Machover, K., Meissner, P. B., and Karp, S. A. (1954). *Personality Through Perception.* New York: Harper.

Yamaguchi, S., Gelfand, M., Mizuno, M., and Zemba, Y. (1997). Illusion of collective control or illusion of personal control: Biased judgment about a chance event in Japan and the U. S. Paper presented at the second confer-ence of the Asian Association of Social Psychology, Kyoto, Japan.

Yang, K. S., and Bond, M. H. (1990). "Exploring implicit personality theories with indigenous or imported constructs: The Chinese case." *Journal of Per-sonality and Social Psychology 58,* 1087–1095.

Yates, J. F., and Curley, S. P. (1996). "Contingency judgment: Primacy effects and attention decrement." *Acta Psychologica 62,* 293–302.

Yates, J. F., Lee, J., and Bush, J. (1997). "General knowledge overconfidence: Cross-national variation." *Organizational Behavior and Human Decision Processes 63,* 138–147.

INDEX

Abelson, Robert, 209
accelerated growth trend, 104–6
acoustic resonance, 21
"action at a distance" principle,
   21–22, 134
acupuncture, 22, 23
advertising, 66–67, 84–85
agency:
  collective, 6, 158
  personal, 1–4, 5, 79, 158
agriculture, 33–34, 36, 43
Airport Site Movie test, 93–95
algebra, xix, 26
amae relationships, 72–73
animism, 84–85
Aristotle:
  analytical approach of, 4, 9, 11–12,
   16, 21, 25, 178–79, 180
  Confucius compared with, xxi, 16,
   29
  ethical system of, 207–8
  intellectual opposition to, 238*n*
Asian Americans, 226–28
Asian vs. Western cognitive processes:
  causality and, 111–19, 210–12,
   226–27
  change as concept in, 174–77
  convergence of, xxi–xxii, 41,
   219–22, 224–29
  homeostatic systems in, xx, 38
  intelligence measurements of,
   212–17

  social structures in, xvii–xviii,
   44–45
  specific examples of, 192–201
assembly line, 82–83
astronomical observations, 7, 21–22
atomism, 80–83, 85, 109
awase style, 76

Bacon, Francis, 203
Bagozzi, Richard, 188
Barnum effect, 185–86
Basseches, Michael, 203–5
behavior:
  alteration of, 207–8, 227–28
  external vs. internal factors in,
   111–20, 124, 152–53
  Fundamental Attribution Error
   (FAE) for, 123–27, 135,
   207–8, 235*n*
  immoral, 198–99
  motivation for, 118–19, 201,
   205–6
  personality and, 119–20, 123–27
  predictability of, 130–33, 135,
   201–2
  rules of, xvi, xvii, 4, 37, 152–53,
   207–8
Beijing University, 86–87, 104–6,
   108, 173–74, 181, 221–22
beliefs:
  cultural differences in, xiv, xvii,
   226–27

beliefs (*cont.*)
  logic of, 171–73
  monotheistic, 199–200
Bellah, Robert, 72
Bell Curve, The (Herrnstein and
    Murray), 216–17
Bersoff, David, 206
biculturalism, 226–29
Big Five personality traits, 122–23
bilingualism, compound vs. coordi-
    nate, 159–62
"bitter drink" experiment, 98–99
Body Adjustment Test, 42
body-soul dichotomy, 154
Bohr, Nils, 225
Bond, Michael, 122
Borges, Jorge Luis, 137
both/and statements, 173–74,
    205–6
Briley, D. A., 183–84
Buddhism, 12, 16–17, 199, 200,
    225
business relationships, 62–65, 66,
    82–83, 98, 194–97, 198

Calvinism, 70, 200
capitalism, 205–6, 219, 222–24
Carnegie Institution, 195
Catholicism, 200
Cattell Culture-Fair Intelligence
    Test, 213–15
causality, 111–17
  ambiguity in, 198
  analytical basis of, 11–12, 210–12
  Asian vs. Western views on,
      111–19, 210–12, 226–27
  binary, 173–74, 205–6
  complexity and, 23–24, 100, 103,
      108–9, 128–29, 134–35, 198,
      209–10
  dispositional factors in, 111–16,
      117, 125–27, 152
  hindsight fallacy in, 130–33
  historical, 127–29, 210, 219–20,
      222
  inference in, xix, 163, 184–85, 202
  models of, 127–30, 131, 134
  physical, 11–12, 21, 36, 109,
      116–17

situational factors in, 111–14,
    116–17, 124–27
ultimate, 179–80
see also logic
chain stores, 83
change:
  in Chinese philosophy, 13–15,
      103, 106–9, 153
  complexity and, 103, 108–9
  contradiction and, 174–76
  cycles of, 103, 106–7, 108, 109, 200
  in dialectical reasoning, 174–77
  in Greek philosophy, 10–12, 103,
      108–9
  linear progression of, 106–8
  in personality, 16, 120
  qualitative vs. quantitative, 204
  stability vs., 45, 102–9, 152
  trends in, 104–8
Cheung, Fanny, 122
ch'i (spirit), 15
children:
  American, 80, 81–82
  Chinese, 57, 87–88, 140–41
  independence of, 57–59
  Indian, 115
  individualism and, 49–50
  Japanese, 55, 57–58, 59, 81–82
  language skills of, 148–52
  math skills of, 153, 188–89
  personal experiences of, 87–88
  relationships of, 50, 57–58,
      150–51, 225
  socialization of, 221
  Western, 57–59
chimpanzees, 225–26
Chinese civilization:
  agriculture in, 33–34
  capitalism and, 223–24
  change as viewed in, 13–15, 103,
      106–9, 153
  children in, 57, 87–88, 140–41
  compromise in, 6–7, 73
  covariation in, 101–2
  Cultural Revolution and, 71, 121
  cultural superiority of, 40
  ecology of, 32, 33–34
  education in, 15–16, 30–31, 72,
      188–89

emotional expression in, 60–61
ethnic homogeneity in, 31–32
family in, 15–16, 71
Greek civilization compared with,
    xix, 1, 6–7, 9, 17, 18–19,
    30–32, 41, 77, 80–81, 108–9,
    137–38, 139, 152–53
holistic world view of, 79, 80–81,
    108–9, 137–39, 152–53
"hundred schools" period of, 6
interdependence in, 77
Japanese civilization compared
    with, 71–72
mathematics in, xix, 19, 26–28
moral values of, 6
philosophy in, 9, 12–19
rationality in, 165, 203
science in, xix, 7, 19, 20–24,
    208–9
self-identity in, 51, 53
social harmony in, 5–8, 19, 32,
    48–49, 54–55, 73–74
technological innovations of, 7
U.S. relationship with, 197–99,
    206
Chinese language, 51, 53, 151–52,
    156, 158–62
Chinese Personality Assessment Sur-
    vey, 122–23
Chiu, Liang-hwang, 140
Choi, Incheol, 119–20, 125, 129,
    131–33, 147, 185–87, 208–9
Choi, Soonja, 152
Chou dynasty, 34
Chou En-lai, 13
Christianity, 69–71, 200, 225
Chuang Tzu, 138
chuan men (make doors a chain), 5
city-states:
    Greek, 30, 39
    Renaissance, 39–40
civilizations, clash of, 219–20
civil rights movement, 206
cognition:
    basic processes of, xiv
    chronological, 127–29, 210,
        219–20, 221
    in females vs. males, 99–100
    formal operations of, 203–4

genetic basis of, 216–17
language and, xix, xxiii, 159–62,
    210–12
materialistic analysis of, 33
modification of, 226–29
normative analysis of, 201–5
objective vs. subjective, 20
schematic model of, 33–37
social influences on, xvii, 39–45,
    57–59
theories of, 85–86, 201–6
universal attributes of, xiii–xv,
    64–65, 192, 201–2
cognitive scientists, xiii, xvi–xvii
Cohen, Dov, 88
complexity, 23–24, 100, 103,
    108–9, 110, 128–29, 134–35,
    198, 209–10
Comte, Auguste, 85
conflict resolution, 194
Confucianism, 6, 7, 12, 15–16, 50,
    103, 195, 199
Confucius, xxi, 7, 15–16, 29, 30,
    106–7, 188
contextual relativism, 204
control:
    achievement of, 96–102
    illusion of, 100–102
corporations, 62, 63–65, 72
correspondence bias, see Funda-
    mental Attribution Error
    (FAE)
cost-benefit analysis, 167
covariation-detection studies,
    95–96, 101–2
Cromer, Alan, 37–38, 209
Cultural Revolution, 71, 121
"culture-fair" tests, 213–15
curiosity, 4, 7–8, 21, 31, 40, 208–9

"dax" experiment, 81–82
Death of a Salesman (Miller), 71
debate, 3, 6–7, 18–19, 20, 25,
    37–38, 194–95, 209, 210–11,
    225
Dershowitz, Zachary, 43–44
Descartes, René, 26
Diamond, Jared, 41
dichotomies, 154–55

Dick and Jane, 49–50
Dien, Dora, 72
diversity (coverage) argument,
    146–47
Doi, Takeo, 72–73
Doris, John, 207
Drerus, 3

Earley, P. Christopher, 98
ecology, 32, 33–34, 38, 45
economics, xv, 42–44, 205–6,
    222–24
education:
    Chinese, 15–16, 30–31, 72,
        188–89
    debate in, 3, 6–7, 18–19, 20, 225
    Greek, 4, 30
    Japanese, 55, 72, 188–89
Educational Testing Service, 215
efficiency, 82–84
egalitarianism, 70
either/or statements, 173–74, 205–6
elders, respect for, 195
Ellsworth, Phoebe, 60
Embedded Figures Test, 42–43,
    228–29
emotions:
    contradictory, 187–88
    expression of, 59–61
    reason vs., 154
empiricism, xiii, 154, 201–2
English language, 88–89, 156,
    158–62
Epidaurus, theater at, 1–2
epistemology, 17, 36–37
equilibrium, 204
erabi style, 75–76
ethics, 207–8
ethnic diversity, 217
ethnocentric thinking, 4
events:
    conflict vs. convergence of,
        219–20, 222
    context of, 127–29, 204
    historical, 127–29, 210, 219–20, 222
    prediction of, 130–33, 135
    relationship between, xix, 108–9,
        175–76
evolution, 154

"face," 71–72
families:
    Chinese, 15–16, 71
    Japanese, 71–72
    Protestant, 69–71
feng shui, 23, 200–201
Fernald, Anne, 150
Fichte, Johann Gottlieb, 176
field dependence, 42–44, 96
Fischhoff, Baruch, 130–31
Five Elements, 15, 138
flying-bird task, 215–16
"focal" fish experiment, 89–92, 118
Ford, Henry, 82–83
formalism, xiv, xvi, xix, 25–26, 37,
    45, 166–67, 188, 190, 203
forms, ideal, 8–9, 156–57
French language, 221
Freud, Sigmund, 85
Fukuyama, Francis, 219, 220
Fundamental Attribution Error
    (FAE), 123–27, 135, 207–8,
    235*n*
Fung, Yu-lan, 21

Gabriel, Shira, 67–68
Galileo Galilei, 21, 22, 41, 178, 180
Garden of Eden, 106, 107
Gardner, Wendi, 67–68
Gelman, Susan, 151–52
Gemeinschaft, 56–57
General Semantics, 238*n*
generic nouns, 151–52, 156–57
Gentner, Dedre, 81–82, 149
geometry, xix, 26, 38, 209
Gesellschaft, 56, 57
goals:
    definition of, 128
    group, 48–49
    personal, 47–48
    properties of, 134–35
God:
    existence of, 179–80
    monotheistic conception of,
        199–200
Golden Mean, 15
Goodman, Nelson, 202
Gopnik, Alison, 152
Graham, Angus, 165

Grand Eunuch, 40
gravity, 21
Greek civilization:
    agriculture in, 33–34, 36
    analytical world view of, 79,
        80–81, 82, 108–9, 137–38,
        139, 152–53, 166–67, 200
    change as viewed in, 10–12, 103,
        108–9
    Chinese civilization compared
        with, xix, 1, 6–7, 9, 17, 18–19,
        30–32, 41, 77, 80–81, 108–9,
        137–38, 139, 152–53
    city-states in, 30, 39
    debate in, 3, 6–7, 18–19, 20, 25,
        37–38
    ecology of, 32, 33–34
    education in, 4, 30
    individualism in, 2–3, 4, 18–19,
        30, 69, 77
    mathematics in, xix, 24–28, 38
    mercantile basis of, 30, 34
    multiculturalism in, 31
    personal agency in, 1–4, 5
    philosophy in, 8–12
    polytheism of, 200
    science in, xix, 7, 20, 21, 37–38
Greek language, 3, 156–57
Gries, Peter Hays, 197, 198
Guns, Germs, and Steel (Diamond),
    41
Gunz, Alex, 88

Hall, Edward T., 50
Hampden-Turner, Charles, 62–65,
    69, 70, 83–84, 197
Han, Jessica, 87–88
Han, Sang-pil, 66–67
Han ethnic group, 31
happiness:
    achievement of, 108
    Chinese ideal of, 5–6
    Greek ideal of, 2–3, 5
Harman, Gilbert, 207
harmony, social, 5–8, 19, 32, 48–49,
    54–55, 73–74
Hayakawa, S. I., 238n
Heath, Shirley Brice, 157
Hedden, Trey, 87, 213

Hegel, Georg Wilhelm Friedrich,
    176
Heider, Fritz, 85
Heine, Steven, 55–56
Heraclitus, 10–11, 29–30
Herrnstein, Richard, 216–17
hindsight fallacy, 130–33
Hinduism, 200
history, 106–8, 127–29, 210,
    219–20, 222
Hofstede, Geert, 62, 70
holism, 17, 79–96, 99–100, 108–9,
    129–30, 137–39, 152–53,
    175–77, 189–90, 210–12
homeostatic systems, xx, 38
Homer, 3
Hong, Ying-yi, 118
Hong Kong, 118, 120–23, 160, 161,
    183–84, 226–27
Hong Kong, University of, 118
horse collar, invention of, 39
human-animal dichotomy, 154
Human Inference (Nisbett and Ross),
    xiv–xv
human rights, 198–99
Hume, David, xiii, 154, 201–2
hunter-gatherers, 43
Huntington, Samuel, 219–20, 222
Hu Shih, 50
hypotheses, 74–75, 133, 209

IBM, 62
I Ching, 14
ideas, marketplace of, 195
identity, law of, 176–77
ideographs, 215
Iliad (Homer), 3
Imai, Mutsumi, 81–82
Indian civilization, 16–17, 26
Indians, East, 114–15
individualism:
    advertising and, 66–67
    alienation and, 225
    characteristics of, 47–48,
        198–99
    children and, 49–50
    collective responsibilities vs., 5, 6,
        15–16, 17, 48–49
    communication and, 60–61

individualism (*cont.*)
  emotional expression and, 59–61
  in Greek civilization, 2–3, 4,
    18–19, 30, 69, 77
  identity and, xvii, 2–3, 4, 18–19,
    48–56
  relationships and, 48–49, 50,
    72–73, 77
  risk assessment and, 98–99
  social values and, 67–68, 221–22
  western progression of, 69–70
Indo-European languages, xxiii, 9,
  156
industrialization, 43
inference, causal, xiv–xv, xix, 163,
  184–85, 202
infinity, concept of, 26–27
in-groups, 51, 52, 98–99
intelligence testing, 212–17
international relations, xvii–xviii,
  197–198
irrational numbers, 24–25
irrigation, 34
Islam, 222
Iyengar, Sheena Sethi, 58–59, 120

Japanese Americans, 227–28
Japanese-Australian "sugar contract"
  case, 66, 197
Japanese civilization:
  advertising and, 84–85
  business culture of, 66, 68, 195,
    197, 198, 222–23
  capitalism and, 222–23
  children in, 55, 57–58, 59, 81–82
  Chinese civilization compared
    with, 71–72
  compromise in, 73
  education in, 55, 72, 188–89
  family in, 71–72
  management in, 195, 198
  U.S. relationship with, 227–28
Japanese language, 51–52, 53, 54,
  150, 156, 158–59
Jefferson, Thomas, 70
jên (benevolence), 51
Ji, Li-jun, 86–87, 95–96, 101,
  104–5, 108, 140–41, 159
Jing, Qicheng, 213

job stability, 63–64
Johnson, Lyndon B., 206
Jones, Edward E., 124–25
Jouvenal, Bertrand de, 102
Judaism, 3, 43–44, 107, 174

Kane, Gordon, 12
Kant, Immanuel, 176
Kennedy, John F., 206
Kieras, David, 212
Kim, Beom Jun, 142, 171–72
Kim, Hee-jung, 54, 210–11
Kitayama, Shinobu, 35, 57, 97, 227–28
knowledge:
  acquisition of, 4, 7–8, 21, 31, 40,
    208–9
  analytic vs. holistic, 17, 79–86, 96,
    99–100, 108–9, 129–30,
    137–39, 152–53, 175–77,
    189–90, 210–12
  organization of, xvi, xix, 23–24,
    134–35, 137–63
  self-, 27–28, 55–56, 76–77
Knowles, Eric, 119
Korean language, 52–53
Koreans, 52–53, 59–60, 66–67, 74,
  119–20, 125–27, 129–33,
  147, 168–70, 172–73,
  186–87, 188, 224
Korzybyski, Alfred, 238*n*
kosmos (magistrate), 3
Kühnen, Ulrich, 228–29
Kyoto University, 89–95

Langer, Ellen, 100–101
language, 155–63
  back-translation of, 88–89
  categorization in, 9, 155–63
  cognition and, xix, xxiii, 159–62,
    210–12
  compound vs. coordinate bilin-
    gualism and, 159–62
  contextual, 51–53, 87, 156–59,
    162–63
  development of, xix, 148–52
  Sapir-Whorf hypothesis of, 159–62
  subject-prominent vs. topic-promi-
    nent, 157–58
  see also specific languages

Lao-tsu, 211
"Lapis Lazuli" (Yeats), 187
law, contract, 66, 196–97
laws, natural, 24
lawyers, 193–94
learning disabilities, 153, 210
Lee, Angela, 67–68
Lee, Fiona, 115–16
Leichtman, Michelle, 87–88
Lepper, Mark, 58–59
Leung, Kwok, 120–21
Lewin, Kurt, 85
Lin, Yutang, 72, 165–66
Liu, Shu-hsien, 165, 203
Lloyd, Geoffrey, 6–7, 20
Locke, John, xiii, 70
Logan, Robert, 11
logic:
    abstract, 8–9, 11, 167–68
    cause and effect in, *see* causality
    compromise and, 183–85, 208–9
    content vs. form in, 165, 203, 205
    contradiction in, xix, 16, 25–26,
        31, 37, 173–77, 180–88,
        191–92, 202, 208–9
    decontextualization in, 25–26,
        167–68, 176–77, 179
    deductive, 202
    dialectical reasoning vs., 27–28,
        37, 45, 174–83, 204–5, 212
    dichotomies in, 154–55
    experience vs., 167–73
    formal, xiv, xvi, xix, 25–26, 37, 45,
        166–67, 188, 190, 203, 205
    identity in, 176–77
    inductive, 139–40, 147, 201–2
    morality vs., 167
    necessary and sufficient conditions
        in, 154–55, 166
    noncontradiction in, 166, 176–77,
        180–85
    plausibility of, 170–73, 181–83,
        237n
    premises in, 168–69
    propositions in, 25, 168–69,
        171–73, 208–9
    rhetorical, 37–38, 166–67, 196
    in science, 74–75, 134–35, 203,
        208–9
    subordinate vs. superordinate cate-
        gories in, 168–69
    of syllogisms, 25–26, 168–70, 203
    truth of, 167, 208
    two-valued, 205–6, 238n
"lottery ticket" experiment,
    100–101
Lu, Gang, 111–14, 129
Luria, Alexander, 85–86
Luther, Martin, 41

McGuire, William, 171
McIlvane, Thomas, 112–14
magistrates, Chinese, 39
magnetism, 21
management, business, 62–65,
    82–83, 98, 194–95, 198, 223
Mao Tse-tung, 175
Markus, Hazel, 35, 54, 57
Marxism, 85
Masuda, Taka, 60, 89–92
mathematics, xix, 19, 24–28, 38,
    153, 188–89, 203
means-ends judgments, 152
medicine, 22, 193, 225
merchant class, 30, 34, 39–40
Merton, Robert, 86
metaphysics, xvii, 35–37
Meyer, David, 212
Michigan, University of, 58, 86–87,
    89–95, 104–6, 108, 173–74,
    177–78, 181, 221–22
Middle Ages, 39, 193
Middle Kingdom, 40
Middle Way, 27, 32, 37, 75, 107,
    176, 177–78, 181, 184–85,
    188, 208–9, 212
Mill, John Stuart, xiii
Miller, Arthur, 71
Miller, Joan, 114–15, 206
mind-body dichotomy, 154
Ming jia (Logicians), 166
Miyamoto, Yuri, 97
modularization, 82–83, 201
modus ponens, 170
Mohists, 166, 167
monotheism, 199–200
moral values, 6, 67–68, 167,
    198–99, 207–8, 221–22

More, Thomas, 106
Morikawa, Hiromi, 150
Morling, Beth, 97
Morris, Michael, 111–14, 118,
    120–21, 183–84
Moser, David, 156–57
Mo-tzu, 26, 29–30, 166, 167
Munro, Donald, 8, 50–51, 138
Murray, Charles, 216–17
music, monophonic vs. polyphonic, 7

Nagashima, Nobuhiro, 166
Nakamura, Hajime, 8, 72
nature:
    complexity of, 23–24, 100, 103,
        108–9, 110, 128–29, 134–35,
        198, 209–10
    definition of, 20, 21, 24, 128
    holistic view of, 17, 79–96,
        108–9, 129–30, 137–39,
        152–53
nature-nurture dichotomy, 154
Needham, Joseph, 18
negotiation, 66, 75–76, 196–97
Newton, Isaac, 41
New Yorker, 83
New York Times, 112, 113
Nissan Corp., 85
Nixon, Richard M., 206
Nobel Prize, 195
Norenzayan, Ara, 119–20, 142, 144,
    171–72
North Korea, 74
nouns, xix, 148–52, 155, 156–57

objects:
    abstraction of, 8–9, 151–52,
        156–57, 199–200
    attributes of, 8–10, 23–24, 36, 80,
        134–35, 137–47, 152–57,
        190, 199–200, 211
    background of, 42–44, 89–92, 93,
        95, 96, 191, 207–8, 228–29
    categorization of, xvi, 4, 9–10, 18,
        19, 23–24, 137–63, 190,
        191–92, 211
    context of, xvi, xix, 10, 21–26, 27,
        36, 42–44, 87, 89–92, 93, 95,

        109, 134–35, 153, 156,
        189–90, 191, 228–29
    discrete, 80–81
    essence of, 9–10, 18–19
    focal, 89–92, 118, 126–27,
        157–58, 191
    individual-class relationship for,
        139–47
    labels for, 150–52
    manipulation of, 79, 96–102
    motion of, 11–12, 116–17
    mutability of, 10–12
    negative matches of, 144–46
    "nonsocial," 87
    part-whole dichotomy for, 138–39
    relationships between, 93–96,
        108–9, 139–47, 162–63,
        189–90
    rules of behavior for, xvi, xvii, xix,
        4, 37, 152–53, 207–8
    substances vs., 80–82, 100,
        138–39
    target, 142–44, 168–69
Odyssey (Homer), 3
Olympic Games, 2
out-groups, 51, 52, 98–99
Oyserman, Daphna, 228–29

Park, Denise, 87, 213
Parmenides, 11
Peng, Kaiping, xv–xvi, 60, 95–96,
    101, 112–14, 116, 118, 119,
    173–74, 177–78, 181, 187,
    197, 198
perception, 79–96
    analytic vs. holistic, xix, 17, 79–96,
        99–100, 108–9, 129–30,
        137–39, 152–53, 175–77,
        189–90, 210–12
    "bottlenecks" in, 212
    control and, 96–102
    economic and social factors in,
        42–44
    priming of, 118–19, 228–29
    of relationships, 108–9, 141–42
    of self, 185–86
perceptual-motor performance, 212
Persians, 4

personality:
  behavior and, 119–20, 123–27
  change in, 16, 120
  Fundamental Attribution Error
    (FAE) for, 123–27, 135,
    207–8, 235n
  harmony in, 122–23
  traits of, 121–27, 207–8
Philosophical Investigations (Wittgen-
    stein), 154–55
philosophy, 8–19
Piaget, Jean, 85, 203–4
place number system, 26
Plato, 8–9, 11, 39, 106, 156–57
polytheism, 200
postformal operations, 204–5
Presbyterianism, 70
probability, 131–33, 171–73
problem-solving, 212–17
progress, 106–8
Protestantism, 69–71, 200
proverbs, 173–74
psychology, 201–5
Puritanism, 70, 106
Pythagoras, 24–25

Raven's Progressive Matrices Test,
    213
reasoning:
  abstract, 8–10, 12, 17–18, 24, 82
  cultural differences in, xiv, xxi,
    111–19, 210–12, 226–27
  dialectical, 27–28, 37, 45,
    174–83, 204–5, 212
  emotional state vs., 154
  historical, 127–29, 210, 219–20,
    222
  logical, see logic
  as process, xiv–xvi, 174–83,
    203–5
  schemas of, 203–4
  universal application of, xiii–xv,
    64–65, 192
rebirth, 200
recall tasks, 89–92
recognition tasks, 89–92
reflective equilibrium, 202
Reformation, 69–70

"regret" formula, 198
relationships:
  amae, 72–73
  business, 62–65, 66, 68, 75–76,
    82–83, 98, 194–98, 222–23
  context of, 61–62
  dyadic (two-person), 72
  employer-employee, 225
  equality in, 48, 49
  event, xix, 108–9, 175–76
  hierarchical, 48, 72–73
  human vs. primate, 225–26
  independent vs. interdependent,
    47–77
  individualism and, 48–49, 50,
    72–73, 77
  linguistic, xix, 148–52
  mother-child, 57–58, 150–51, 225
  object, 93–96, 108–9, 139–47,
    162–63, 189–90
  obligations in, 49
  perception of, 108–9, 141–42
  reciprocal, 204
  rules for, 64–66
relativism, 202–3, 204
religious wars, 199–200
Renaissance, 39–40
Republic (Plato), 106
resonance, 17, 21, 27–28, 138
retrieval cues, 87
rhetoric:
  legal, 74, 75
  linear, 196
  logic of, 37–38, 166–67, 196
  scientific, 74–75, 209
Riegel, Klaus, 203–5
risk assessment, 98–99
Rod and Frame Test, 42, 101–2
Rosemont, Henry, 5
Ross, Lee, xiv–xv, 123
running-bird task, 215–16
Russell, Bertrand, 203

St. Luke, 4
Sanchez-Burks, Jeffrey, 59–60
Sapir, Edward, 159
scholē (leisure), 4
Schwarz, Norbert, 86–87

science:
  in Chinese civilization, xix, 7, 19,
    20–24, 208–9
  in Greek civilization, xix, 7, 20,
    21, 37–38
  hypotheses in, 74–75, 209
  intellectual confrontation in, 195
  logic in, 74–75, 134–35, 203,
    208–9
Sears Roebuck Company, 82
self:
  attributes of, 50–51, 53–54
  collective agency of, 6, 158
  common perceptions of, 185–86
  individualism and, xvii, 2–3, 4,
    18–19, 48–56
  knowledge of, 27–28, 50–56,
    61–62, 68, 76–77
  linguistic definition of, 51–53
  non-Western vs. Western, 48–56
  personal agency of, 1–4, 5, 79, 158
  social context of, 27–28, 50–53,
    61–62, 68
  uniqueness of, 53–56
self-esteem, 53–56, 68
self-transforming systems, 204
Sethi (Iyengar), Sheena, 58–59, 120
Seven Coincidences Board, 215–16
Shang dynasty, 34
Shavitt, Sharon, 66–67
Shintoism, 84
Simonson, I., 183–84
sin, 200
Singapore, 161
single-motive fallacy, 205–6
skill development, 55–56
Skinner, B. F., 86
Smith, Adam, 205–6
Smith, Edward E., 142, 147
Smith, Linda, 151
social structures:
  in Chinese vs. Greek civilizations,
    30–32
  cognition and, xvii, 39–45, 57–59
  ecological influences on, 34–35,
    38, 45
  economic influences on, 30, 34, 38
  folk metaphysics and, 35–37

harmony of, 5–8, 19, 32, 48–49,
    54–55, 73–74
  moral values in, 67–68, 221–22
  self-knowledge and, 27–28,
    50–53, 61–62, 68
society:
  autocratic, 2, 3
  collectivist, 57, 67–68
  democratic, 1–2, 39, 209, 219,
    223
  egalitarian, 107
  hierarchical, 6, 15, 31–32, 34, 77
  high-context vs. low-context, 50
  oligarchic, 39, 223
  utopian, 106–7
Socrates, 30
South Korea, 74, 224
Sowell, Thomas, 70
spatial skills, 215–16
statistics, xv, 209, 215
status, achieved vs. ascribed, 63–64
Stevenson, Harold, 221
Stich, Stephen, 202
Su, Yanji, 104–5, 108
surgery, 22, 193
syllogisms, 25–26, 168–70, 203

tai chi, 225
Taiwan, 161, 162
Taoism, 12, 13–15, 16, 27, 51, 84,
    103, 138, 175
Tao Te Ching, 14, 138
Tardif, Twila, 149, 151–52
Tönnies, Ferdinand, 56–57
Trompenaars, Alfons, 62–65, 69, 70,
    82–83, 197
tunnel vision, 89, 96

University of Hong Kong, 118
University of Michigan, 58, 86–87,
    89–95, 104–6, 108, 173–74,
    177–78, 181, 221–22
utopias, 106–7

verbs, xix, 148–52, 155
virtue ethics, 207–8
Vranas, Peter, 207
Vygotsky, Lev, 85–86

Wang, Qi, 87–88
Watanabe, Masako, 127–28
Weber, Max, 85
Westernization, 220–22, 224
Whorf, Benjamin, 159
Witkin, Herman, 42, 96
Wittgenstein, Ludwig, 102, 154–55
women:
   control as perceived by, 102
   holistic view held by, 99–100
Wong, Nancy, 188
World Journal, 112, 113

Yamaguchi, Susumu, 98–99
Yang, Kuo-shu, 122
Yeats, William Butler, 187
Yi, Youjae, 188
yin-yang principle, 13–14, 15, 16,
   27
yoga, 225

Zeno, 11
zero, concept of, 26–27
Zhang, Zhiyong, 140, 159
Zuozhuan, 7

ABOUT THE AUTHOR

Richard E. Nisbett, Ph.D., has taught psychology at Yale University and currently teaches at the University of Michigan, where he is the Theodore M. Newcomb Distinguished University Professor. He has received the Distinguished Scientific Contribution Award of the American Psychological Association, the William James Fellow Award of the American Psychological Society, and, in 2002, a John Simon Guggenheim Foundation Fellowship. He is the author and editor of several university press titles. He lives in Ann Arbor, Michigan.